Finance and Control for Construction

In a climate where margins are tight and profitability depends absolutely on successful financial management, it is crucial that managers in the construction industry are aware of the way projects are financed and controlled from the initiation by the client through the final handover of the building, to the financial implications of running and maintaining costs.

Covering the whole lifespan of a building, this book explains the theory and practice of financial management from development, through the design, procurement and post-contract processes.

Chris March has a wealth of practical experience in both the construction industry and teaching students. His down-to-earth approach and mixture of theory and real-life evidence from personal experience make him the ideal guide to learning successful financial control and management.

Chris March is a graduate from Manchester University. He worked for John Laing Construction and later for John Laing Concrete where he became Factory Manager. On entering higher education he worked in both the UK and Hong Kong before joining the University of Salford becoming Senior Lecturer and then the Dean of the Faculty of the Environment. He is a former winner of the Council for Higher Education Construction Industry Partnership Award for Innovation.

Finance and Control
for Construction

Chris March

Spon Press
an imprint of Taylor & Francis

LONDON AND NEW YORK

First published 2009
by Taylor & Francis
2 Park Square, Milton Park, Abingdon, Oxon OX14 4RN

Simultaneously published in the USA and Canada
by Taylor & Francis
270 Madison Ave, New York, NY 10016

Taylor & Francis is an imprint of the Taylor & Francis Group, an informa business

© 2009 Chris March

Typeset in Sabon by
HWA Text and Data Management, London
Printed and bound in Great Britain by
TJ International Ltd., Padstow, Cornwall

British Library Cataloguing in Publication Data
A catalogue record for this book is available from the British Library

Library of Congress Cataloging-in-Publication Data
March, Chris.
   Finance and control for construction / Chris March.
      p. cm.
   Includes bibliographical references and index.
   1. Building – Cost control. 2. Construction industry – Finance. I. Title.
TH438.15.M37 2009
690.068´1–dc22   2008037921

ISBN10: 0-415-37114-7 (hbk)
ISBN10: 0-415-37115-5 (pbk)
ISBN10: 0-203-92804-0 (ebk)

ISBN13: 978–0–415–37114–8 (hbk)
ISBN13: 978–0–415–37115–5 (pbk)
ISBN13: 978–0–203–92804–2 (ebk)

Contents

Figures

Tables

Introduction

This book is one of three closely related texts, *Finance and Control for Construction*, *Operations Management for Construction* and *Business Organisation for Construction*; the reason for writing these books is the increasing awareness of the shortage of new texts covering the whole range of construction management. There are plenty of good recent texts appropriate for final-year and postgraduate students primarily, but they tend to be subject-specific and assume a certain level of knowledge from the reader. It also means students find these texts costly and tend to rely on the library for access. (The research selectivity exercises have encouraged authors to write books based on their research, for which credit has been given in the assessment, whereas none has been given to those writing textbooks.) The purpose of these three books is an attempt to give students the management vocabulary and understanding to derive greater value from these specialist texts.

The original intention was to write these three books with construction management undergraduate students in mind, but as the project developed it became clear that much of the subject matter was appropriate for all of the construction disciplines. Recently, the industry, having too few good construction graduates, has turned to recruiting non-cognate degree holders. Many of these graduates will study on Masters courses in construction management. These texts are ideally suited to them as background reading to give a broad base of understanding about the industry.

Rather than having a large number of references and bibliographies at the end of each chapter, generally I have limited these to a few well-established texts, some referenced in more than one chapter, so the reader is directed to only a few for further and more in-depth reading on the subject. The chapters vary in length considerably depending on the amount of information I believe is relevant.

The aim of this book is to offer the reader the opportunity to track the control of finance through each stage of the building process. It is not the intention to enter the area of accountancy, but rather to give an understanding of how the various parties become involved in the process, their contributions and the reasons for the various elements of cost controls. The text considers four main areas: first, deciding whether or not to build in the first place; second, controlling the costs of the design; third, establishing a cost upon which there is a basis for a contract with a constructor; and finally how the constructor controls costs of the construction process.

Interwoven in the text are other issues, for example, in the design processes, environment and sustainability. Where some background is felt necessary to aid the understanding of financial control, then this has also been included, such as a brief resumé of the development process.

The two related books – *Operations Management for Construction* and *Business Organisation for Construction* – are concerned respectively with the management of site operations and facilities management, and the running of the business. There is inevitably overlap in all three books, so I have cross-referenced from one book to another and within each subject, with the hope of aiding readers.

On a personal note, I believe that there is no definitive way of managing and, as Mike Stoney, the managing director of Laing, used to say to students, 'Don't copy me, it may not suit your personality, but watch and listen to other successful managers and pinch the bits from them that suit you'. I totally agree and, for what it is worth, I have added some other people's comments and thoughts that have influenced my way of thinking over the years.

My headmaster Albert Sackett – who taught me to assume everything was wrong until I could prove otherwise – would set an essay on, say, 'define the difference between wit and humour'. After he marked it he would sit you in front of the class and debate with you your answer. After convincing you he was right and you were wrong, he would then reverse role and argue back the opposite way.

Godfrey Bradman, chairman of Bradman Management Services, reinforced my views from school of not simply accepting anything you are told, but also having the ability to ask the right questions – usually simple ones such as 'why not?'.

My father taught me to accept that failure was a fact of life and not to hide the failure, but to accept it, admit it and get on with life having learnt from the experience. I have also found that by admitting the failure, the 'punishment' is always less than if it is kept hidden. When in the precast factory it was always easier to advise the site manager that the load of components was going to be late or not delivered that day than await the angry phone call demanding to know what had happened. It also made sense because, although disappointed, they had more time to rearrange their own schedule of work.

I learned another piece of advice from Dorothy Lee, retired Deputy Director of Social Services in Hong Kong and advisor for the Caritas-operated Kai Tak East Vietnamese refugee camp. After many weeks of hard work on a self-build solution for refugees, the camp was no longer required. She said, 'I know you will be disappointed, but remember you have grown a little more as a result'.

John Ridgway, explorer and outside activities course organiser, had at his School of Adventure, based in Sutherland, Scotland, the adage of positive thinking, self-reliance and leaving people and things better than you found them. He also made a very clear impression on me of the importance, when in charge, to have the ability to stand outside the circle and allocate tasks without becoming too closely involved.

Don Stradling, the personnel director of Laing and senior negotiator with the Federation of Civil Engineers Employees gave the very simple piece of advice that 'you should always keep the moral high ground'. How right he is. It is surprising the number of people who don't, and when confronted with one who does they almost invariably fail in the negotiation. It also results in having respect from those they have contact with, as they believe in your integrity and accept what you say is meant in an honourable way.

Finally, Dennis Bate, member of the main board of Bovis Lend Lease told me that throughout his life – from leaving school at age sixteen to become an apprentice joiner – he strove to do whatever he did to the best of his ability – and do it better than anybody else.

I wish to acknowledge the support and help given by so many in putting together these three books. From the construction industry, staff from Bovis Lend Lease, Laing O'Rourke and Interserve in particular, who have spent many hours discussing issues and giving advice. It was at Plymouth University that the idea to produce these texts was formulated and where colleagues gave me encouragement to commence; then at Coventry University, not only was this continued, but doors were always open whenever I wished to consult on an idea or problem. My years at Salford, from 1987 onwards, when we started the Construction Management degree, were of great significance in developing ideas on the needs of construction management students. This would not have been possible without the assistance and guidance from my colleagues there at the time, especially Tony Hills, John Hinks and Andy Turner, as well as the many supporting contractors always on hand to give advice, ideas and permit access to other colleagues in their organisations. Finally, to my wife Margaret who had to suffer many hours on her own whilst I locked myself away in the study, but who never ceased to give her support and encouragement.

Chris March

1

The development process
The main participators

Before considering the economics of development, it is important to realise the range of different people who can become involved in the process. Not all those described will always be involved as it depends upon the type and scale of the development. In certain instances some of those identified are described in more detail than others, especially if their role extends into the construction phase as well as the development. There can also be overlap in the roles, whereby a developer could employ several specialists, or for smaller development projects, a more generalist service provider. A large developer might have some of the personnel employed in its own organisation supplemented by outside specialists as required.

1.1 Landowners

All developments need either virgin, previously used or reclaimed land, all of which will be owned by an individual, group, corporate body or public authority. Without their permission, unless compulsorily purchased, a development cannot take place. Landowners may either initiate the development or be approached by others to release land for the development to take place.

1.1.1 Traditional landowners

These include

- the Church of England
- landed aristocracy
- landed gentry
- the Crown Estate.

Jointly, the above landowners own a significant amount of the national land bank. For example, the Duchy of Cornwall comprises some 57,000 hectares; the Duke of Westminster owns an even greater amount of land, including 120 hectares of Mayfair and Belgravia in London. The Church of England owns 5,000 hectares of agricultural land and some 16,000 churches, of which 12,000 are listed as being of special architectural or historic interest; some of these come onto the market as 'redundant churches'. The Crown Estate is part of the hereditary possessions of the Sovereign the profit from which is paid into the Exchequer. It incorporates an urban estate which includes significant London holdings in Regent Street, Regent's Park and St James's Park, as well as just under 5,000 hectares of agricultural land and extensive UK marine assets.

Generally, in these cases, the motives for ownership are more than just capital return and include social, political and ideological reasons, often based on tradition and heritage.

1.1.2 Industrial and commercial landowners

These include:

- farmers
- nanufacturers
- industrialists
- extractive industries
- retailers and a variety of other service providers, such as banks
- public authorities, such as central, local and nationalised industries that also own land.

The value of this land to the landowner is usually incidental to its main purpose. Primarily, farmers use the land to produce crops or rear livestock. Whilst the initial purchase price of the land is a contribution to the overheads the farmer incurs, the fact the land may be an appreciating asset doesn't affect the price of the produce as this is determined by other forces and only comes in to play if the farm is to be sold. The only other occasion when the value of the land is likely to be taken into account is if the farmer wishes to borrow money against it. The same logic similarly can apply to the others listed above.

Whilst the land is fundamental to their business, these landowners do not perceive it as an asset in the same way as those who purchase land with a view to capitalising on it. This means that the economic reasons for releasing land are not always obvious to them. As a result, they may be reluctant to co-operate with a potential developer unless a good case can be made to them. An example of how the land value could be of importance is if it is of high value it may be profitable to sell the land and with the monies released, re-locate elsewhere and build an up-to-date, purpose-made, energy-efficient premises, thereby reducing overheads and becoming more profitable.

1.1.3 Financial landowners

These people and groups purchase land, with or without property built on it, as an investment and therefore are more likely to co-operate with a developer and include:

- builders
- property developers
- pension funds
- insurance companies.

The latter two have invested from between 5 and 15 per cent of their funds in property investment over the last 20 years and have a significant portfolio of property.

Those purchasing land are playing a long-term game: anticipating healthy profits in the future, knowing that the land, even if not appreciating in the short term, will almost certainty be saleable. Land banks are crucial for the long-term viability of residential developers if they wish to maintain a flow of completed properties. This can include green belt land, held in hopes that the Secretary of State will release it for building as a result of pressure in the housing market. It can be derelict areas in inner cities were there is a potential for future development, and many areas are being regenerated due to the demand of having living accommodation closer to one's place of work. In essence, the purchaser is looking for potential opportunity.

1.2 Private developers

Private-sector developers come in a variety of sizes, from one person to multinationals that may be publicly quoted on the stock exchange. Their purpose is to make a direct financial profit from the process of development. They operate as either traders who sell the property they develop or as investors who make their money from renting the property and from an appreciating asset. They may also develop the property for their own occupation and use. Examples of the latter include banks and building societies, which have a considerable property portfolio.

Traditionally, those involved in residential development were usually traders who build to sell. However in recent years housing associations have had a significant impact on the housing stock and they build with a view to let. The other major change in development has been with the advent of Private Finance Initiatives (PFI), which fall into two main categories. The first is facilities management: when a new building with a specific purpose, such as a hospital or school is built and the developer then rents the property back to the user and contracts to maintain the building over a period of years, usually 25 to 35 years. The second is when an existing building is to be refurbished, and the whole or part of the building is to be handed back to the client and is supported with a facilities management contract. The Treasury Building in London is an example of being able to use half of the floor area for development other than Treasury business. In essence, the contract was to find alternative accommodation for the Treasury employees during the refurbishment, redesign the interior in a more effective way so that on returning they could function in half the footprint of the building and then the developer could use the other half for other purposes that were acceptable to the client. The developer could then make a profit from this development as well as from the facilities management provided for the client.

In the total development process it is the developers and those that provide the finance for the project who take the greatest risk. Post-war history is littered with examples of those who have made great profits when the rents and property values have risen, and failures when the bottom has dropped out of the market. It is extremely difficult to predict market trends with certainty as fluctuations in the world economy can occur as a result of unpredictable events. The high increase in oil prices in 1973, the miners' strike during the Thatcher era, the dismantling of the Berlin Wall and the reunification of Germany, the Exchange Rate Mechanism crisis, September 11th, a deterioration or improvement in major economies such as in America, China and Japan, and the credit crunch of 2008 are examples of such events.

1.3 Public-sector and government agencies

There was a time when the central government carried out or supported the local authorities in doing a considerable amount of direct development. Socialist administrations more than Tory – but in recent years, the policies of both parties – have converged to carry out little direct development. If the government requires work to be done, increasingly it is through the PFI route using private finance rather than Treasury monies.

Local government can be come involved in developments within their boundaries, but it depends upon the interests of their community. If the area is derelict or run down, developments that improve local amenities, are likely to be welcome. On the other hand if the area is highly sought after, the community may not wish development as they may consider this to be a loss of amenity.

Some interesting work has been done in partnership with the private sector. An example of this is the Barnsley Metropolitan Borough Council and Costain Group partnership where the local authority enlisted the help of the construction company to plan and improve the area as well as seek financial support and investment, for which they obtained a fee. Whilst the company was also able to tender for construction work they were not always successful as systems were put in place to ensure the local authority got best value for money (see section 8.6).

The more active authorities can act as a catalyst by supplying the land, which they lease to the property developer or by investing in the infrastructure. Salford Quays is an example of the latter. The authority, with the aid of central funding, carried out the refurbishment of the harbour structures themselves and installed good quality paved roads to attract developers, and at the same time set a standard of quality which would encourage the developers to follow. For example, the materials used for the car park areas provided by the developer are generally of similar quality to those used by the local authority.

In an attempt to foster development, especially in inner cities, the government has produced urban regeneration initiatives administered through various government agencies, such as:

- urban development corporations
- English Partnerships
- The Welsh Development Agency
- Scottish Enterprise and local enterprise companies.

These groups see themselves as assisting developers with land assembly, site reclamation, provision of infrastructure and financial grants.

Other initiatives aimed at attracting occupiers with financial incentives include Enterprise Zones and Regional Selective Assistance. There are also Regeneration Zones specified by the European Commission, which also give financial incentive for development.

1.4 Planners

The most significant change to the planning system in the UK occurred with the enactment of the Town and Country Planning Act 1947 from which all subsequent legislation has been developed.

Both politicians and professional planners control planning. The former, in local and central governments, are responsible for approving or refusing development plans in accordance with the policy laid down by them. They will be advised and guided by the professional planners who also administer the system on behalf of the politicians. This normally means that unless the development is controversial, the politicians will accept the planner's recommendations. The basis for determining permission is laid down in statute and a variety of central government policy guidance notes. A local authority is obliged to work to these, but will determine its own local policy normally using the local development plans which will have been approved by central government.

There are two main reasons for planning. The first is to prevent development that is undesirable, for example, if it is out of character with its surroundings aesthetically, is a use not compatible with other users in the area such as positioning industrial premises in the middle of a residential area, or places excessive demands on the existing infrastructure. The second is to encourage development where it is appropriate. For example, if an area is declining it is important to encourage investment to breathe new life into the area either by improving the overall amenities or by attracting new industry and commerce.

In practice, there are a lot of opportunities for interpretation of the local plans and developers may employ their own planning consultants to assist in negotiating with the planners. There is also the issue of 'planning gain' when developers may be expected to, or offer to, provide something extra as part of the agreement. This may be an improvement in the local roads, landscaping or other amenity for the local community.

There is no national standard as to what development is acceptable. This depends on the local area and whether or not there is a need to encourage development. Clearly, if the authority wishes to attract investment, it is likely to require lower standards, and if it wishes to slow down or deter development it will impose higher standards.

There has been an increasing trend by developers to use planning appeal procedures as a result of conflict between the developers and the planners and sometimes because of intervention by the politicians who have ignored their planners' advice.

1.5   Financial institutions

Financial institutions usually refer to pension funds, insurance companies, clearing and merchant banks (both UK and foreign). Building societies also provide finance, although most of this funding is for the private residential market. How the financial institutions provide finance is discussed in more detail in Chapter 2.

1.6   Agents

Commercial or estate agents (in the case of residential) may see a potential opportunity for development and bring together the key players in the process. Increasingly, they are likely to be commissioned by the developer to find suitable sites or properties for development, redevelopment or refurbishment.

The knowledge these players bring to the process at the early stages is an awareness of the current market. For example, what the current demand is and what current rents and prices are; essential to assess the project's viability. Linked to this is advice on potential occupiers, the mix of tenants and their likely requirements in terms of design, layout and space. However, it should be noted this is a specialist area and if the agent does not specialise the developer would commission market research independently for more detailed information. Commercial and estate agents will also be aware of the planning implications of the project in the location being considered and be able to advise. Depending upon the experience of the client they will assist or advise on how to obtain finance for the project.

They are commonly employed as the selling or letting agents. Here their role is to advertise, negotiate with potential future occupants or owners, and in the case of tenants, manage the tenant agreements on behalf of the developer. This entails the collection and processing of rents, ensuring the property is being maintained as agreed and administrating any dilapidations schedules at the end of the tenancy.

They can also be employed by the landowner to protect their interests when approached by the developer. The landowner may not be aware of the development potential and could receive a lower price than it is worth. If on the other hand the landowner wishes to sell land or property they may not know its likely value and will employ an agent to advise. A simple example of this is when a homeowner wishes to sell their property and employs an estate agent to value the property and act as agent in the selling process.

1.7 Building contractors

Traditionally, building contractors were brought in to construct the building after the project had been financed and designed. However in recent years, many different forms of procurement and contract forms have been developed and used. Typical examples are design and build, management contracting, construction management and PFI (see Chapter 8). Developers now often engage a contractor or project manager to manage the whole process including financing, the design and build of the project, and the facilities management. For example, the last two developers competing for the Treasury Building PFI contract employed MACE and Bovis Lend Lease as the project managers during the bidding process.

Developers, especially retailers such as Marks & Spencer and Tesco, sometimes develop partnerships with contractors to build their outlets. There are advantages of doing this for both parties. The developer can negotiate competitive rates with the contractor and use its expertise to improve the performance of the design and specification. The contractor has guaranteed work over a period of time and can designate certain key personnel to these projects because of the continuity of work, which in turn benefits the developer. The contractor can develop partnerships with sub-contractors and suppliers that also can produce similar benefits and again the developer can benefit as a result. In essence it enhances the possibility of good supply chain management (see *Operations Management for Construction*, Chapter 7).

Contractors can also be the developers, especially in the provision of residential accommodation. Some have built up a significant land bank over many decades that they release for their own speculative development. Such organisations will also purchase run-down premises, notably in inner cities and either upgrade them or convert them into private residential developments.

1.8 Project design and development team

Once the decision is made to seriously consider putting a development project into practice, a team of experts needs to be put together. Most developers do not have the skills or expertise in-house to carry out major development so they employ professionals/specialists/experts to advise them at the various stages of the process. These may include some or all of the following:

1.8.1 The architect

Anyone can call himself a builder, surveyor or engineer, unless prefaced by the term Chartered, but in terms of designing buildings the title 'architect' is protected by law. Architects must be qualified and registered with the Architects Registration Board.

Traditionally, the architect was responsible for the design of the building, obtaining planning permission and supervising the construction operation on behalf of the client. However, whilst this still occurs, on most major projects, management of the whole process is carried out by others, the role of the architect being confined to the design of the building only. On some projects specialist space planners may support their work (see section 1.8.2).

It is a misconception to believe the architect's function is solely to produce an aesthetically pleasing building or a 'work of art'. Whilst this is part of the role, the talent they bring to the process is considerably more than this. There are instances where it is important to make a dramatic statement such as The Lowry, the millennium landmark project for the arts at Salford Quays, Canary Wharf in London and The Guggenheim Museum, Bilbao. Normally the design of the building has to blend in with adjacent buildings or those in the locality using appropriate building materials, style, form and colour.

However, hidden behind the building's façade is where the architect's other key contribution can be found: making the building work for the owners and occupiers. The first stage of this process is the designing of the footprint of the building and its orientation. Fundamental strategic decisions are made at this stage especially those concerned with environment, in particular the amount of energy required to run the building. If as much natural daylight and ventilation are to be used as possible, this limits the depths of rooms measured from the outer wall. The orientation of the building can affect such issues as solar gain or noise pollution. The way in which the architect creates working/living and circulation spaces within this footprint are of crucial importance if the building is to function properly and efficiently. They are also involved in the specification and selection of components and materials used for the construction of the building.

1.8.2 Space planner

Space planning is one of the tasks architects provide as part of the overall design service. However, specialist planners are often employed, especially on large projects, to organise spaces such as open-plan offices, to accommodate the client's needs. This will often involve establishing the client's requirements either through discussion or to an already prescribed brief. In the process of their work, they may also demonstrate that the space available is unsatisfactory to satisfy the client's work practices.

1.8.3 Structural engineer

The prime role of the structural engineer is to ensure the building will not collapse or deform significantly when in use. Secondary roles are to achieve this as economically as possible, provide a solution that satisfies the architect's design needs and to accommodate building services.

The engineer has to take into account the final use of the building and the resultant loads that will be placed on the structure through to the foundations. The architect will have produced floor plans and elevations as well as information on suggested floor-to-floor and floor-to-ceiling dimensions. The engineer has then to design solutions that satisfy these requirements, if possible. If not then the architect may well have to reconsider their design and agree a compromise. The materials for the structural frame may have also been determined for a variety of reasons such as speed of construction, the amount of repetition and client's specifications.

A considerable amount of money is spent on the foundations. However, if the project is being fast-tracked, much of the information may not be available at the time of the design, resulting in the engineer being cautious and producing a more expensive solution than is finally necessary.

During the construction phase, the structural engineer will inspect and monitor the project to ensure that the structural work complies with the specifications stipulated. This will be done in close liaison with the contractor's engineering staff who will also discuss any structural design issues that occur as the project progresses.

1.8.4 Building services engineer

The costs of the building services as a percentage of the final cost of the building work can vary from 25 to over 60 per cent, depending on the nature of the building's use. Mostly hidden from view it comprises a labyrinth of pipes, ducts and wires travelling both vertically and horizontally throughout the building to deliver services which include water, power, lighting, communication, heating and ventilation. Much of these will require a plant of some significant size and weight, which will have to be supported structurally and given a place in the building. Added to these issues could be means of transporting people and goods such as escalators, travellators and lifts.

Clearly, the input of a building services engineer into an integrated design process is crucial if a satisfactory solution for the functioning of the completed building, its maintenance and its buildability during construction is to occur. They often are able to contribute to the environmental debate, especially in terms of energy consumption by proffering ideas and suggestions on how strategic and detailed design decisions can be modified to reduce the building's reliance on certain services. This could result in cost savings by reducing the size and capacity of the plant, not to mention reducing the weight on the structure and possibly the floor-to-ceiling dimensions in which ducts have to be accommodated.

The selection of plant and equipment to run the building services are also important, as they will have to be replaced and maintained during the life of the building. The cheapest may not be the most economical, looking at the long-term financial implications as it may breakdown more often, require more frequent servicing and have to be replaced in a shorter time than a more expensive equivalent (see Chapter 6).

1.8.5 Quantity surveyor

The quantity surveyor (QS) is concerned with the cost of the construction of the project and these days also needs to consider the future life and cost of maintaining the building. The developer needs an indication very early on what the building on completion will cost. This allows for a decision to be made as to whether or not to progress with the project, either in its current form or in some modified way. In the latter case, this may require producing a less ambitious solution or, more likely, cutting back on the standard of finishing and site works to make the necessary savings. The amount of information available will determine the technique used for producing the cost plan. Using these methods permits the surveyor to produce the costs of alternative design solutions relatively rapidly.

A key role for the QS is to draw up the contract documents that will be used for the selection of a contractor and subsequent cost control of the project during construction. Traditionally, this was the bills of quantities and one of the standard forms of contract such as JCT05 Standard Building

Contract. However, only approximately 30 per cent of contracts now use the bills of quantities and other less cumbersome forms of control have been developed. This has enabled the time from initial design to contractor selection to be reduced considerably.

Finally, in liaison with the contractor's QS, the amount paid by the client to the contractor for work done each month is agreed, any variations from the original design and assumptions are validated and the final account settled at the end of the contract. Whilst not part of the design and development team, the contractor's QS's role is discussed here because they represent the contractor's close link with the process during construction.

The contractor's QS is there to look after the contractor's financial interest as the contract progresses and is completed. As indicated before, each month they agree the measure of work that has been done and the amount of materials stored on site for use in the contract so the contractor can obtain payment for the work done each month. This is important, as it is from this money that the contractor is able to pay its sub-contractors and suppliers. During the contract they record and notify any variations to the work that has been tendered for and any legitimate delays beyond the contractor's control and do similarly with the sub-contractors and suppliers who themselves may have claims.

They will also advise the contractor's management about any contractual implications in the tender documents during the estimating process. Finally they work to ensure that at the end of the contract the contractor is paid for everything they are entitled to.

1.8.6 Environmental consultant

Any development – whether it is a refurbishment, a new building or infrastructure – will have an impact on the environment both during the construction and during the life of the building. Some contracts, such as the Mass Transit Railway in Hong Kong, not only take a long time to complete but also extend through large areas of urban communities. The impact on the community therefore can be considerable, not just during the construction process, but afterwards in, for example, the levels of noise generated through a 24-hour operating day. The environmental consultant carries out environmental impact assessments and advises on measures that can be taken to alleviate the problems.

Environmental consultants need to be brought in at the earliest stage of the design process to impact on the strategic design decisions, such as the footprint and orientation of the building otherwise their contribution to the environmental debate will be minimal. As the design process moves to the detail, the consultant advises on the selection of materials. Issues for consideration are the embodied energy of the product or material, pollution implications during manufacture and eventual disposal on demolition, whether it is sourced from sustainable sources, its long-term durability, and the environmental implications of maintenance, such as cleaning. All of these in the context of ensuring the material or component satisfies the technical criteria of the design for its application and is within sensible cost parameters (see Chapter 6).

1.8.7 Building surveyor

Traditionally, building surveyors were concerned with the state of the existing building, analysis of the causes of the defects and producing solutions to rectify the situation. This has placed them in an ideal position to be involved in development of existing property, be it maintenance or refurbishment. When a property is being considered for refurbishment, which may also mean a change in use, such as Victorian warehouses being converted into residential units, the property has to be initially surveyed to establish its condition but also to see if it is practical to convert it to its new use. Design solutions can be drawn up for consideration and the process then continued in a similar manner to any new build development. Resulting from their expertise, building surveyors are sometimes employed as the project manager for the whole project.

1.8.8 Lawyers

Lawyers are involved at various stages of the project, and on major developments this may involve using specialists from different practices. This starts with the acquisition of the land and or existing property that may require negotiation with several interested parties if the development is to cover several landowners, some of whom may only wish to lease, whereas others may wish to sell.

   If there is a need to go through the planning appeal process, solicitors and barristers may be used to present the developer's case at the enquiry.

   Often there has to be put in place legal agreements with those funding the development and contracts drawn up between the developer and the professionals concerned with the design and management of the project. These contracts may be significantly different from the standard contracts, such as the JCT form. It also may be necessary, for example, in the case of government work, to ensure that all those who are given access to certain material sign the Official Secrets Act.

Finally, contracts have to be drawn up between the developer and those occupiers who are either leasing or purchasing a part or the entirety of the development.

1.8.9 Accountants

On a regular basis, governments change the tax and VAT regulations, usually in the annual budget and subsequent announcements. Regional development funding instigated by the government and the European Commission also change from time to time. Many of these changes are very complex and need specialist accountants to understand the ramifications. They can also be used to advise on financing agreements and on the structure of partnerships as well as on methods of raising finance.

1.8.10 Valuation surveyors

Valuation surveyors are usually brought in very early in the process to assess the viability of a project on the proposed land. They may be used to seek out different potential sites for the development and conduct valuation surveys to establish the most suitable for the developer's needs. They will provide such information as the cost of the land and the likely returns the developer could expect to make by leasing or selling. They would be expected to provide knowledge on taxation. Much of what they do can be provided in more specialist forms by some of the participants previously described, but they provide an extremely useful service in their own right. They can be employed during the formative design process to value different solutions and uses for the proposed development.

1.8.11 Facilities management consultant

There are many large projects, especially PFI, which require the property to be maintained after it has been built or refurbished, for a period up to 25 or 35 years as part of the overall contract. The provision can include all types of maintenance such as a cleaning service, repairs and planned maintenance. The latter involves both the maintenance of existing plant and the fabric of the building as well as replacing components at the end of their useful life. It would cover day-to-day repairs such as damage and failure, from light bulbs to lift breakdowns. Checking of fire alarms and systems would be included, and in certain circumstances might include security, car parks and the postal services within and to and from the building. The cost implications of providing this all-inclusive service are immense and it is essential that the correct advice be given at the design stage as incorrect decisions made then could have a major knock-on effect in the future (see Chapter 6).

1.8.12 Project management

Finally, someone has to manage the process, including those previously named in this chapter. This may be the developer itself or it may be delegated to one of the major players. Traditionally, this would have been the architect, but in recent years it is more likely that a management contractor or project management organisation would be employed. For a refurbishment project a building surveyor could be employed.

1.9 Objectors

Objection to development has almost become part of our national life. These can be issues of relatively minor significance, except to the objectors, to issues of regional and national importance. Opposition can be costly to the developer because of delays incurred, the cost of planning enquiries, increases in standards of materials and components, and, at worst, having to abandon the project. However with foresight much of this can be anticipated and contained. There are several categories of objectors.

- Self-interested neighbours of the proposed development, often referred to as NIMBYs (not in my back yard). They can organise and significantly obstruct the progress of the development proposals and have been known to successfully cause delay, as well as cause the project to be abandoned.
- Environmental protestors who come from outside the local area. They can be very persistent in their protest, and some take illegal action by occupying the site, putting themselves at danger by tunnelling, tree dwelling and moving to anywhere where it is difficult and dangerous for them to be evicted. They are, of course, seeking high levels of publicity and aim to draw attention to the environmental issues that concern them. Note it is usually the contractor, who is only the artisan in the process, who usually gets the bad publicity rather than the developer or planner. It is an interesting discussion point to formulate ways of dealing with these situations if they arise.
- The professional, permanent bodies at local and national levels, such as the Victorian and Georgian societies, Friends of the Earth and

Greenpeace. There are also the official quangos such as The Nature Conservancy Council and English Heritage. All of these organisations are well organised, with a high level of devoted supporters. Because of their knowledge and experience of the planning and development processes, they have the capability to cause problems especially to an inadequately thought-out project which has neglected issues with which these kinds of objectors are concerned.

Whilst not objectors, as meant in this section, it should not be forgotten that politicians can overturn a development project if they believe it is not in the public interest or if it could affect their likelihood of re-election. At the national level the Secretary of State has powers to overrule either the development proposal or those who are objecting to it. The planners employed by local government can recommend a development should not take place although they can be overruled by the politicians.

1.10 Occupiers

Occupiers are often unknown at the beginning of the development process, and their needs are not always fully researched, but should be. Development is often tuned to suit the needs of the financial institutions rather than the occupier. There is a growing realisation that this has to change, and in shopping malls, for example, the development may well not commence until some of the bigger retailers are signed up. If the occupier is known at the beginning of the design process then they can become involved in some of the key decisions so their requirements can be met.

Whereas retailers often see the space around them and that which they occupy as part of the shopping experience, commercial occupiers are more likely to regard the building as an overhead incidental to their business. They do not always appreciate the importance the building's design, layout and provision of services in contributing to the successful running of their operation, especially in terms of accommodating change in the way they work.

Occupiers are demanding much more flexible lease arrangements to enable them to react to changing needs. The financial institutions would prefer long-term lease agreements for obvious reasons, but are having to come to terms with the changing marketplace.

References

Ashworth, A. (2002) *Pre-Contract Studies, Development Economics, Tendering and Estimating*, 2nd edn, Blackwell Publishing.

Baum, A. (2000) *Freeman's Guide to the Property Industry*, Freeman's Press.

Baum, A. and Mackmin, D. (1989) *The Income Approach to Property Valuation*, 3rd edn, Routledge

Cadman, D. and Topping, R. (1995) *Property Development*, 4th edn, E&FN Spon.

Curwell, S., Fox, B., Greenberg, M. and March, C. (2001) *Hazardous Building Materials – A Guide to the Selection of Environmentally Responsible Materials*, 2nd edn, E&FN Spon.

Darlow, C. (ed.) (1993) *Valuation and Development Appraisal*, 2nd edn, The Estates Gazette Ltd.

Harvey, J. (2000) *Urban Land Economics*, 5th edn, Macmillan Press.

Lupton, S. (2001) *RIBA Handbook of Practice Management*, RIBA.

Ratcliffe, J. and Stubbs, M. (2001) *Urban Planning and Real Estate Development*, UCL Press.

Rougvie, A. (1988) *Project Evaluation and Development*, Mitchell.

Scarrett, D.T. (2000) *Property Valuation*, E&FN Spon.

Taylor, N.P. (1991) *Development Site Evaluation*, Macmillan.

2

Sources of finance

This is not meant as a definitive list of sources but to give the reader an indication of the kinds of organisations that can be considered and to indicate for which type of development they are most likely to be appropriate.

The finance required for development falls into two basic categories. First, 'development finance' or 'short-term money' that is needed to cover all the costs incurred in purchasing the land, the development, design and construction processes. This may be paid back shortly after completion if the development is sold on and a profit made. Second, 'funding' or 'long-term money' that is used to cover the costs incurred when holding onto the completed development as an investment. This money is repaid using the revenue generated from renting the property. The monies for each may not come from the same source.

Some developments are financed entirely from the developer's own capital, but this is the exception rather than the rule. Most developers will go to one or more of the financial institutions to seek funding as this spreads the risk in the event of the project failing. However, if they have surplus cash it may be in their interest to invest it in property especially if the expected returns from rental growth are good.

2.1 Insurance companies and pension funds

The pension funds and insurance companies take a long-term view, as they need to achieve capital growth to pay out the agreements made with pensioners and policyholders. They tend to be more cautious and conservative than other lenders as a result of their responsibility and in recent years have reduced the percentage of property investment in their portfolio in preference to equity shares, but they are still major investors in property. However, since the fall of the stock market in 2002/3 they have moved from equity to safer havens and property is often considered to be a safer long-term investment. They will have a clearly defined set of criteria that have to be satisfied before being prepared to invest. As a result they expect to have significant control on the ways monies are to be spent, the quality of the building, its location and financial status of the tenants. The outcome of this is that developers are inclined to produce development schemes that satisfy the financial institutions rather than that of the future occupants.

They tend to invest only in large-scale projects especially those in prestigious sites and occasionally become the developer taking on the risk. Generally their aim is to minimise risk and maximise future yields; yield being the annual income received from the asset expressed as a percentage of its capital cost or value (see section 3.2.14).

2.2 Banks

The banks are approached if the development is not acceptable by the insurance and pension companies or the developer cannot or will not provide the necessary financial guarantees. The banks will use the property asset as security for the loan. It is attractive as it is a large identifiable asset with a resale value.

2.2.1 Clearing banks

The clearing banks normally offer short- and medium-term loans, although this is not always the case. Much depends on such factors as how buoyant the market appears, current government policy and the track record of the developer. They are more flexible in their approach to loans than the insurance companies, especially for refurbishment and development of older buildings. They also take an interest in the business of the prospective occupier and of the property development company, as well as the property itself. If the loan is high, the bank may secure an equity stake in the project as they are exposed to a greater risk. Residential developers tend to go to the clearing banks as their requirements are only for a short-term loan.

2.2.2 Merchant banks

Merchant banks are more entrepreneurial than the clearing banks and for that reason are likely to take a greater risk. At the same time they will tailor-make a financial package for the individual developer. They may also divide the funding into smaller packages and obtain funding for each package from other banks. The cost of borrowing from them is usually higher than from the other main sources of funding.

2.3 Private person

Any individual who purchases a property is an investor and many in the UK have entered the housing market as a means of increasing their wealth and as a hedge against inflation, although this is not always a successful short-term strategy if the market value falls and the owner enters negative equity. Without these types of investors the speculative housing market would not exist in its current form and housing development projects could well be different. The majority of these investors purchase the property they live in, but others purchase with a view to refurbish and sell on at a profit, whilst others accumulate property and earn rental income.

2.4 Building societies

Building societies lend primarily to the domestic property market because they rely on the income from the mortgages to supply finance on a regular repayment that enables money to be lent to others.

2.5 Government and EC funding

Generally speaking government has too many pulls on its purse strings to give out money for development, and this is in part why they support the Private Finance Initiatives. However, there is funding available for certain initiatives especially if they wish to encourage development in an area that might otherwise be unattractive to developers, usually in the form of subsidies or reduced rates for a fixed period. They may contribute to projects, which could have national significance such as buildings for a proposed Olympic bid. They also indirectly fund projects with money raised from the National Lottery, such as the Millennium Dome, the new Wembley Stadium and the Lowry Building.

The Department of Trade and Industry may offer capital grants towards the costs of factories, and local authorities may also be willing to contribute to the development of local amenities and tourism.

The European Commission also assists, almost invariably, in the funding of projects in the more deprived areas of the EU when regions are given Objective 1 or 2 status with the aim to improve the economy of the area. This is a very complex area of funding which is continually changing. Most of the funding is directed for a particular purpose and not normally directly to the developer. The developer can, however, work in partnership with the eventual end user to attract funds to enable the project to come to fruition.

Valuation and development appraisal

3.1 Introduction

A simple definition of development valuation 'involves the calculation of what can be achieved for a development once completed and let, less what it costs to create' (Ratcliffe *et al.* 1996).

The developer needs to establish whether or not the project is likely to be viable. At this stage the information available is limited and this section aims to indicate the various factors and the methods that can be adopted to inform the decision as to whether or not to proceed. This section concentrates on valuation required for new projects and not the valuation of existing property.

The costs involved in a development include the land, design, construction and all the fees to the others involved in the process as outlined in Chapter 1. There are various questions the developer needs answers to depending upon the information available. These are:

- If the development costs and profit margins are known, what is the maximum cost of land that can be absorbed to make the project viable?
- How much profit can be made if the costs of the land and construction are known?
- How much rental income is required to justify the development?
- If the land value and profit are known what is the most that can be spent on the construction costs?

3.2 Cost elements

There are many cost elements the valuation surveyor needs to put a value against in order to calculate answers to the questions raised above. Following are typical examples to be considered, although not all will be used in every case.

3.2.1 Cost of the land

This may be the price, after negotiation, that the landowner is asking for the land, or the price the developer is able to offer after taking into account the cost of the development and the returns expected after selling or leasing and the profit required. The value of the land to the developer is not necessarily the asking price but the difference between the cost of the building to be built on the land and the market price of the finished development including the land. If this value is less than the market price for the land, then the development is not feasible.

Sometimes the client owns the land. In the case of government and local authorities the land cost may not be included in the overall budget calculations. In others, such as speculative housing, the developer will wish to pass on the current value of the land to the purchaser to release funds to purchase more land for further work. The value of the land will be related to supply and demand and in certain cases may only be available on lease – often 99 years.

The value of the land will be affected by several factors such as:

- Its geographic location. For example, land in the centre of London will be some of the most expensive in the world, whereas that in the Highlands of Scotland may be very cheap.
- Its proximity to the transport links, especially to road and rail, but also air; the latter being good for business but not so for residential.
- Its proximity to other development and local facilities.
- The topography of the land. Is it level or hilly and does it have a high or low water table?
- The level of contamination. This can be a very expensive to remedy costing many millions of pounds, as shown in section 3.2.4.
- Public rights of way through the land that may have to be moved or could result in possible protests.
- Restrictions on its use either required by the seller or as a result of planning restrictions. The former can be very restrictive in the case of leasehold land. Some residential leasehold agreements restrict the

parking of caravans, erecting of sheds and greenhouses and, in some cases, satellite television dishes. However, the owner of the property has the right to buy the freehold after three years.

- The state of the national and regional economy.

The price of the undeveloped land equals the value of the development less the building costs and developer's profits. Since the seller of the land can also do this calculation they can set the land price accordingly. This means that contrary to common belief, it is not the cost of the land that pushes up the price of the building, it is the other way around. It is the market price of the development, which pulls up, or pushes down, the cost of the land as the developer can only get what the marketplace permits.

3.2.2 Legal and professional fees in acquiring the land

At the feasibility stage the developer requires to know what the developed value of the land is, as described above. To obtain this a valuation surveyor will be engaged (usually a Chartered General Practice Surveyor). Once the sums have been completed and it is agreed the land is to be acquired, fees for land agents and lawyers fees will ensue. The valuation surveyor may also be asked to value different plots of land for the development. For example, if it has been decided to relocate a government department to an area, the valuation surveyor will look for plots of land that can accommodate the development and will produce a report outlining the costs of the development (building and land) on each of the plots along with the advantages and disadvantages of each location.

3.2.3 Gross building size and lettable area

This is a function of the area of the site which can be practically built on and local planning policies which may place restrictions on the density of the development, the maximum heights of buildings and so on. There also may be conditions laid down over means of access, numbers of car parking spaces permitted, landscaping, protected trees and views. Taking all these into account will determine the size of the building and the maximum amount of usable space.

3.2.4 Construction costs

The construction cost is the greatest outlay of all at a time when no revenue is being generated. It is important that this figure is as accurate as possible. This is a function of how far the design process has progressed. The means of estimating with incomplete information is discussed in Chapter 4. There are certain elements that need to be looked at in particular as the costs of these can be much higher than predicted if not based on some detailed investigation and analysis. Examples of these are:

- *Demolition of existing structures.* The cost of demolition existing properties will vary dependant upon:
  - the value of the materials in the existing building in terms of the likelihood of recycling
  - the volume of material to be disposed of, because of landfill charges
  - the distance from the tip
  - the ease or otherwise of dismantling
  - whether or not it is in a busy urban area, because dust and noise become a greater issue, as do difficulties associated with transport and access.
- *Decontamination of land.* This can be very expensive depending on the area and depth and type of pollution. It is not the intention to discuss methods of dealing with contamination here, but readers should ensure they are familiar with the various techniques available. Contamination is usually found in brownfield sites where there has been industrial use. It can be as a result of petrol spillage or, in certain parts of the country, from mining activities, most notably coal. The latter can pollute watercourses and water tables. Another major source of contamination occurs when building on landfill sites. It is said that the decontamination costs of the Millennium Dome site was in of the order of £200 million.
- *Archaeological finds.* If an archaeological find is discovered on the land, then work must stop to permit an archaeological dig to take place. The time this takes depends on the significance of the find. It may be that the finds are just recorded and the construction can then continue, or in extreme cases, the finds are so significant as for example in Shakespeare's Rose Theatre in London, the development may have to stop totally. Whilst finds cannot always be predicted the likelihood of finds in certain cities such as Chester are more probable than in others. Searching archives for information could be worth the effort in these cases.

- *Previous building work below ground.* It is not unusual when constructing on previously used ground to find a whole array of potential problems which have not been recorded. Examples of these include cellars and redundant underground services the statutory exteriorities have no record of or are in a different place than thought. The more detailed the search, the more likely it is that these problems can be anticipated thereby reducing delays and associated costs.
- *Ground conditions.* The type of sub-soil and level of the water table can have significant impact on the costs of the foundations. The ground below the building has to support the weight and the activities that occur within the building. Generally speaking the weaker the soil the more expensive the foundations will be. Undetected variances in the ground type may delay the construction work whilst redesigning takes place. Hence the more detailed knowledge there is about the ground conditions, the clearer and more accurate the costs will be.
- *Building services.* As has already been indicated in section 1.8.4, the costs of the provision of the building services can approach 60 per cent of the cost of the completed building. This means it is often the largest cost element of the building. Thus careful scrutiny is required to ensure an effective integrated solution is produced which takes account of the capital costs of installation and the running costs when the building is in use. It cannot be stressed enough the importance of considering this aspect of the design as early as possible, especially in terms of the strategic design decisions affecting energy consumption.

The total cost of the building is not just the cost of the materials, labour and plant used to construct it, but also the contractor's site and head office overheads, profit and all the temporary works, such as scaffolding, formwork and soil support.

3.2.5 Professional fees

These fees are usually expressed as a percentage of the total costs of construction and vary from 10 per cent (if the building is simple or has repetitive building components) to as high as 17 per cent (on very complex refurbishment projects). In practice, the majority are around 12.5 per cent to 14.5 per cent. These fees can be found in the recommended fee structures by the appropriate professional institutions and the Royal Institute of British Architects (RIBA) *Handbook of Practice Management*. Whilst it is useful to use this figures for development valuation, when it comes to engaging the profession it is also possible to negotiate a fixed fee. It is argued by some that, other than the professionalism of the person engaged, there is no incentive to produce a cheaper solution if one is paid as a percentage of the final cost of the construction work.

3.2.6 Development and construction period

Throughout the development and construction processes, costs accrue which have to be financed. It is therefore essential to know how long these will take so the necessary funding can be put in place. Whilst on the surface it would be reasonable to expect the time to construct the building to be longer than that of the development, this is not always the case. Much depends upon the complexity of the project, the time spent on the planning processes – especially if it goes to an appeal – whether there are several landlords involved, the financing options which depend on the current economic climate, and the purpose and location of the development.

The process of obtaining planning permission can be relatively simple and inexpensive, whereas if it goes to public enquiry it can be drawn out and expensive. In extreme cases such as the proposed King's Cross development in London, it may even have to go before parliament. There can be considerable costs incurred in these processes, in terms of the fees charged by the consultants and the delays inherent in the system.

3.2.7 Costs associated with facilities management

If the project is to include a facilities management contract, the developer will also need to look at the capital costs of providing materials and equipment otherwise provided by the prospective client, including the costs of maintaining and replacement. Examples are:

- *Furnishings and fittings.* These include furniture, carpets, blinds and curtains.
- *Equipment used for the business as distinct from running of the building.* These include kitchen equipment and computer facilities.
- *Costs of managing, running and maintaining the building including energy and waste disposal.* If the project is to be sold on this does not apply. However if the building is leased in part or in total, all of these will have to be costed and paid for by the developer. The building will need maintaining and repairing as parts and components wear out. Decoration is required from time to time and the building will have to be kept clean inside and outside.

- *Rates and insurance.* Rates bills may be passed down to the tenant, but if there is not full occupancy then this proportion of the rates will have to be borne by the developer. The building structure and fabric will have to be insured against damage and fire.
- *Security.* The capital cost of security systems has to be considered as part of the capital cost of the building as does the manpower required to supervise and control the security of the building. In some cases, depending upon the use of the building and the implications of multi-occupancy, this can be an unexpectedly high sum since it can be required 24 hours a day, every day of the year.

3.2.8 Disposal costs

The three types of agency agreement most likely to be used when selling or letting a development are:

- *Sole agency* is an agreement with a single firm. The expected fee would be between 1.5 per cent and 3 per cent of the agreed selling price, or approximately 10 per cent of the annual rental value.
- *Joint agency* is when two agents are instructed by the developer to jointly sell or let the property. This type of agreement is used when, for example, a mixed development may require specialist knowledge by the agent or if the property is being advertised nationally. The developer will have to pay up to 1.5 per cent more than the normal fee.
- *Sub-agency* is when the primary agent may employ another agent to assist in the marketing of part of the development because of the other's specialist expertise. In this case fees are likely to be similar to a sole agency agreement.

3.2.9 Cost of finance

As indicated in Chapter 2 the developer will obtain finance from a variety of sources so as to distribute the risk. In return the developer will be charged interest and this has to be costed into the development calculations. This is a complex issue outside the remit of this text, however it is important the reader understands the issues involved. Finance has to be raised for the cost of the land, the design and construction processes and the period from handover of the building to the receipt of the first income from rents or from selling the completed development. All of these include professional fees as discussed before. The finance required is needed at stages over the development, as these different expenditures have to be covered as they occur.

The simplest way to estimate the finance for construction is to either take half the interest rate over the total construction period, or the full interest rate over half the building period as demonstrated below in section 3.3.2, the residual method. Both methods give different figures. To be more accurate, the costs of construction can be broken down into monthly expenditure and the interest calculation conducted on the amounts of finance raised as it occurs, as shown in section 3.3.3, the discounted cash-flow analysis. Compound interest is usually applied to the finance acquired for the purchase of the land, normally over the entire development period.

The rate of interest paid depends upon the source of finance available to the developer, but for these calculations it is normal to apply a figure a few points above the current base rate.

3.2.10 Rental income

The calculation of rental income is complicated by the fact that it is obtained over several years in which time interest rates can change, demand may vary, not to mention inflationary influences. It is probably the most important cost item in the calculation and yet will rely heavily on experience and prediction rather than hard evidence. It could be argued the most successful developers are those who have a natural ability to get this right.

The evidence is most likely to be used is comparisons with rents obtained on similar properties in the area, adjusted to take account of the nature and location of the property, car parking provision, available floor area and the sizes and shapes of accommodation in the development. The issue is further complicated by having to take into account the running costs of the development. For example, an older property may have higher heating and lighting bills than a new, purpose-built one. They can also have different natural daylight and ventilation characteristics and standards of insulation.

It is also necessary to assess the likely speed of obtaining near to or full occupancy, as this will also affect the revenue stream. Developments with pre-letting agreements with tenants increase the certainty of the calculation. Often in major retail developments, development does not go ahead until some of the major high-street retailers are signed up.

3.2.11 Administrative costs of leasing

If the building is leased to one or more tenants, then rents have to be collected or chased, dilapidation schedules have to be produced and, when tenants leave, the building has to be surveyed and comparisons of its state made against the agreed dilapidation schedules. It then has to be re-advertised and let.

3.2.12 Contingencies

This is a percentage allowance made to cover for any unforeseen circumstances. Typical examples of uncertainty at various stages of the development are unforeseen ground conditions affecting the cost of the foundations and the costs of resulting delays caused by redesign, costs of planning gain and special needs of major tenants. How much is allowed for depends to a large extent on the quality of the information available at the time of the development costs calculation. The more accurate the information then the more certainty and less contingency provision is required. Typical figures are 5 per cent on the construction costs or 3 per cent on the gross development costs used. On refurbishment work this figure will normally be higher because of increased uncertainty. Rather than having a contingency item, some developers increase their expected profit by about 2 per cent.

3.2.13 Required profit

Any development carries with it a potential risk and the amount of return developers expect will take this into account. However, a figure of between 15 and 17 per cent is generally the minimum return a developer would expect on the cost of the development.

3.2.14 Yield

A term commonly used is 'yield' and is an important concept. In essence it means the higher the level of earnings the investors obtain from their investment, the higher the yield and the more attractive the investment. So a 15 per cent investment on £100 will pay £15 per annum, whereas a 5 per cent will only produce £5 per annum. However a word of caution: usually the higher the yield the greater the risk. The developer has to take account of this risk when making the decision as to whether or not to proceed with the project.

3.3 Development valuation techniques

There are various methods of evaluating the viability of a development, including the comparative method, the residual method and the discounted cash-flow analysis. There are other methods in use and an interested reader may refer to such texts as Scarrett(1996) and Ratcliffe *et al.* (1996).

3.3.1 The comparative method

This is a simple method by which the valuer looks at similar properties in the area where the value is known, for example, if it has just been sold. Then, taking account of any differences, the valuer makes a value assessment. Estate agents selling residential properties use this method. The differences will include its architectural design and appearance, location, state of repair, quality of internal finishing, such as bathrooms and kitchens, and any other additions made. This method is effective when the market is stable, but more difficult to use when the market is volatile and house prices are rising or falling almost daily, which is why it is essential when selling a property to obtain several valuations.

3.3.2 The residual method

This method has its limitations, but it does give a quick indication as to whether or not the development might be viable. It should only be used for a quick approximate answer. There have been serious criticisms of the method by the Lands Tribunal, and its use as a technique for valuation in the property boom of the 1970s also brought it into disrepute. The method is not precise enough because in practice expenditure is assessed during the construction phase at monthly intervals, and in the case of the design, fees are often paid at the end of each phase of work in line with the RIBA plan of work (section 4.1). Its use can produce inaccurate forecasts affecting the developer's prospective profit. It can be a useful tool for the developer when considering several alternative projects in eliminating the clearly non-profitable ventures.

In essence, the residual method is based on a simple equation:

Residual value = Gross development value − (Costs + Profit)

As indicated in the Introduction the developer will most likely wish to find out how much he can afford to pay for the land and still make the profit required; or if the cost of the land and building works are known, establish the likely profit from the development. The figures used in the examples below are symbolic rather than true to life, but are used to demonstrate in Tables 3.1 and 3.2 the calculation in two different examples.

Establishing the value of the land

The total floor area of the proposed office building is 3000m² of which 2400m² is lettable, the rest being circulation areas, etc. The initial development period is estimated to be 12 months followed by a 10-month construction phase and an allowance of 4 months to let the property. Construction costs are estimated to be £1000 per m². It is expected to obtain rents of £200 per m² on the lettable space. Finance has been arranged at 12 per cent per annum and a 6 per cent yield is predicted. The developer wants to make 15 per cent profits on the capital value. See Table 3.1.

Table 3.1 Establishing the value of the land

Net development value		£	£
Estimated rental value per annum (pa)	2400m² @ £200	480,000	
Year's purchase (YP) in perpetuity at 6%		16.67	
Estimated gross development value or capital value			8,001,600
Development costs			
Building costs	3000m² @ £800	2,400,000	
Cost of finance on building costs (interest charges)			
(12% pa for construction period on half cost of building)	2.4m × 0.5 × 0.12 × 10/12	120,000	
(12% pa for letting period on full cost of the building)	2.4m × 1 × 0.12 × 4/12	96,000	
Professional fees	12.5% of building costs	300,000	
Cost of finance on professional fees (interest charges)			
(12% pa for construction period on half cost of fees)	300k × 0.5 × 0.12 × 10/12	15,000	
(12% pa for letting period on full cost of the fees)	300k × 1 × 0.12 × 4/12	12,000	
Subtotal		**2,943,000**	

continued...

Table 3.1   continued

Contingency 5% on costs including interest	2,943k × 0.05	147,150
Promotion and marketing	estimate	60,000
Agents fees – 10% of annual rentable value	480k × 0.1	48,000
Sale costs if property sold when fully let		
(3% of capital value)	8,001,600 × 0.03	240,048
Subtotal		**495,198**
Net development costs		3,438,198
Developer's profit 15% on net development costs		535,730
Total development costs		**3,973,928**
Residue to buy land	8,001,600–3,973,928	4,027,672
Residual land value		
Let land value = v	v	
Cost of acquisition @ 2.5%	1.025v	
Cost of finance for land (interest charges)	1.025v × 0.12 × 26/12 = 1.64v	
(12% per annum for 26 months)		
Profit on total land cost @ 15%	1.64v × 1.15 = 1.88	
The price developer can afford to pay for land	4,027,672/1.88	2,142,000

Note interest added – not compounded for simplicity

Establishing the likely level of profit

Using the same information but making an assumption that the land value is £2,500,000, the question posed is what profit will the developer make? See Table 3.2.

Table 3.2 Establishing the likely level of profit

Net development value		£	£
Estimated rental value per annum	2400m² @ £200	480,000	
Years purchase (YP) in perpetuity at 6%		16.67	
Estimated gross development value or capital value			8,001,600
Development costs			
Building costs	3000m² @ £800	2,400,000	
Cost of finance on building costs (interest charges)			
(12% pa for construction period on half cost of building)	2.4m × 0.5 × 0.12 × 10/12	120,000	
(12% pa for letting period on full cost of the building)	2.4m × 1 × 0.12 × 4/12	96,000	
Professional fees	12.5% of building costs	300,000	
Cost of finance on professional fees (interest charges)			
(12% pa for construction period on half cost of fees)	300k × 0.5 × 0.12 × 10/12	15,000	
(12% pa for letting period on full cost of the fees)	300k × 1 × 0.12 × 4/12	12,000	
Subtotal		**2,943,000**	
Contingency 5% on costs including interest	2,943k × 0.05	147,150	
Promotion and marketing	estimate	60,000	
Agents fees – 10% of annual rentable value	480k × 0.1	48,000	
Sale costs if property sold when fully let			
(3% of capital value)	8,001,600 × 0.03	240,048	
Subtotal		**495,198**	
Net development costs		3,438,198	
Land cost		2,500,000	
Acquisition cost @ 2.5%	2,500,000 × 0.025	62,500	

continued...

Table 3.2 continued

Cost of finance on land (interest charges)		
(12% pa for development, construction and letting period)	2,562,500 × 1 × 0.12 × 26/12	664,250
Total development costs		6,664,948
Profit	8,001,600–6,664,948	1,336,752
Developer's percentage profit	(1,336,752 × 100)/8,001,600	16.60%

3.3.3 Discounted cash-flow analysis

Described as a method of evaluating an investment by looking at projected cash flows and taking into account the consideration of the time value of money (i.e. interest and depreciation). Whilst still not completely accurate, this method more truly reflects the actual expenditure incurred by the developer. Its weakness is that it is still vulnerable to any delays that might occur and if expenditure varies from that predicted, the applied interest rates can distort the final picture.

Taking a simple example:

- the cost of the land is £400,000
- the fees for acquiring the land is £20,000
- the building costs are £2,000,000
- the design team fees are £300,000
- the agency fees are £60,000
- a contingency of 5 per cent is added to the building and design costs
- the development is sold for £4,000,000
- the present value (PV) of £1 is 6 per cent
- the design period is 6 months
- the construction period is 18 months
- the development will be sold during the following 6 months.

Table 3.3 demonstrates six-month periods for simplicity, but could in reality be broken down further into either three-month or one-month periods. The figures in brackets are in debit and the others in credit.

As a learning exercise, this information can be entered into a standard spread sheet, then by altering the land costs, the building costs or the PV of

Table 3.3 Cash flow table

Item/months	0	6	12	18	24
Land costs	(400,000)				
Purchase costs	(20,000)				
Building costs		(600,000)	(700,000)	(600,000)	
Design team fees	(100,000)	(50,000)	(50,000)	(50,000)	(50,000)
Agency fees					(60,000)
Contingency 5%	(26,000)	(32,500)	(37,500)	(32,000)	(2,500)
Sale proceeds					4,000,000
Cash flow per period	(546,000)	(682,500)	(787,500)	(682,000)	3,987,500
Cash flow cumulative	(546,000)	(1,228,500)	(2,016,000)	(2,698,000)	(1,389,500)
Present value (PV) of £1 @ 6%	1	0.971	0.943	0.916	0.890
Net present value (NPV)	(546,000)	(662,707)	(742,612)	(624,712)	3,548,875
Cumulative NPV	(546,000)	(1,208707)	(1,951,319	(2,576031)	972,844

£1 by 1 per cent increments, and then different combinations of all three, it will be seen the effect uncertainty can have on the final cash flow sum.

The net present value (NPV), as used in this example, is what the sum of money will be worth at that point in time compared with the present time because it has been discounted, in this case at 6 per cent.

3.3.4 Comparison of residual method and discounted cash flow

There is no doubt there is a role for both methods. The residual method is the more appropriate for an initial appraisal as it provides a more immediate view on the project's viability and remains the most common in practice despite reservations of the Land Tribunal. The discounted cash flow method allows for more fine-tuning, but both have their limitations because of the problem of ensuring the data is accurate (see Table 3.4).

3.4 Risk analysis and sensitivity analysis

The problem with any method of valuation is the number of variables that are involved, most of which are based on estimates. Add to that the fact that

Table 3.4   Disadvantages and advantages of the two methods

Residual method Advantages	Residual method Disadvantages
It is simple to use	Land Tribunal has been critical of the method
Can be used for all sizes of developments	The time scale of the project and the finance is very approximate
The residual can be the profit if the site price is already fixed	A small changes in one of the factors considered can change residual out of all proportion
In simple terms it mimics the market approach, i.e. Gross Development Value – Costs = Existing Value	Many assumptions have to be made on the input data

Discounted cash flow Advantages	Discounted cash flow Disadvantages
More scientific and there is greater analysis	It is considered too academic by some
Detailed analysis of expenditure and income	Can produce over-analysis
A sensitivity analysis can be carried out	Increases the opportunity for input errors
It demonstrates the cash-flow requirements	At early stages of development there is insufficient firm data leading to inaccuracy
Can include possible inflation influences on larger schemes	
Can more readily assess the contribution of the various elements to the scheme	

applying interest rates can be done in different ways, it can be seen there is a significant opportunity for error to occur in the calculations. The purpose of this section is to comment on the potential for error, but not to enter into statistical probability calculations. If the reader wishes to develop this further, several texts can assist, notably Ratcliffe *et al.* (1996).

- *Land costs.* Sometimes the land costs are known because the developer owns the land, but if not then an offer has to be made. If the land is on the market a value will have been assessed by the vendor and the purchaser can then negotiate accordingly. However, if the land is not on the market, then its price and the time taken to purchase the land the land will be less predictable, as the vendor is usually in a stronger position during the negotiation.

- *Purchasing fees*. These are normally a percentage of the cost of the land and therefore the risk is directly proportional.
- *Usable area of the building*. The topography, shape of the site and planning restrictions can all affect the footprint and height of the building that can be built on the site. This in turn may determine the amount of circulation space required, which has to be deducted from the overall floor area to establish the usable area for a specific use. It should be noted that the developer might not have decided what this use might be so this can be significant. Knowledge of the future use and a certain amount of design work will increase the accuracy of this assessment.
- *Cost of the building*. The accuracy of this depends on how far the design has progressed, the quality of the site survey and the type of contract being negotiated with the building contractor. It is possible to agree a fixed price with the contractor taking the risk. For example, the Lowry on Salford Quays had to be designed and built to a fixed budget, as this was all that was available. However, this is unusual and normally the developer would be expected to take part of the risk and extra costs might be incurred as a result of ground difficulties, archaeological finds and other problems.
- *Professional fees*. These can either be a fixed fee or a percentage of the cost of the building, in which case the risk is proportional.
- *Time for design, building and letting*. Any delays which occur during any or all of these processes has a direct effect on the costs of the interest on the finance obtained for the project, but in times of high inflation or shortages of the particular type of property, it has been known to be lucrative for the developer. This would be a very risky strategy plan.
- *Rental value*. Establishing the likely rental value is one of the most sensitive factors. One is predicting the rent that might be collected a considerable time into the future. Since this is a function of supply and demand at the time, there are many unpredictable factors that can come into play, notably world and national economics that can be affected dramatically by unforeseen events, such as September 11th, the SARS outbreak, the credit crunch, or war. This is why developers aim to obtain major tenants in advance of completion and even before commencement of the development. This means a hedge against a changing market place and also reduces the risk of not letting smaller parts of the development, especially in the retail sector, as potential small tenants know that the larger ones will attract customers.

- *Finances for the project*. Before the development starts, all the finance for the design, construction and letting periods will be in place. This can be on either a variable interest rate, leaving the developer exposed to fluctuations, or on a fixed rate. However, in both cases there may be penalties if the project is delayed and the loan period has to be extended.
- *Investment yield*. This is determined by the marketplace and carries the same uncertainties as rental value unless the project is pre-sold.
- *Marketing and selling costs*. The costs of the agent's fees are expressed as a percentage of the selling price of the building or the annual rental fee and are therefore linked to any extra costs incurred in either of these. The developer may also spend extra money on promotion especially if the project is not being let as quickly as anticipated.

It is important the developer looks at all these factors and analyses the risks involved and does not rely entirely on instincts and experience, although these two issues should not be ignored. Since each of the factors outlined carries different potential risks, the range of probability needs to be considered and statistical analysis applied. This is referred to as a sensitivity analysis.

3.5 Mathematics of valuation

3.5.1 Formulae

There are several formulae needed to calculate how money invested will change over the years as a result of interest made or lost. This is important for life-cycle costing calculations and discounted flow calculations used for property valuations. If the reader requires a reference to find out how to derive the formula, see Chapter 10 in Kelly and Male (1993) or Baum and Mackmin (1989).

3.5.2 Compound interest or amount of £1

If money is invested over a number of years it will earn interest at the end of the first year. This interest will be added on and at the end of the second year this new sum will be used to calculate the interest for the second year and so on. The formula for this is:

Compound interest $= (£ \text{ invested} + \text{interest})^{\text{number of years}}$

or for every pound invested

$$CI = (1 + i)^n$$

So if £1000 were invested at 4 per cent over 5 years, the calculation would be as follows:

$$1000(1+0.4)^5 = 1000(1.04)^5 = £1220$$

Therefore, £1000 invested would yield £1220 after 5 years. Since this formula is the basis for the others the derivation is shown here:

Let the interest per annum on £1 be i.
The amount at the end of one year will be $(1 + i)$
The amount at the end of 2 years will be
$(1 + i) + i(1 + i) = 1 + 2i + i^2 = (1 + i)^2$
The amount after 3 years will be
$(1 + i)^2 + i(1 +i)^2 = 1 + 3i + 3i^2 + i^3 = (1 + i)^3$
Hence after n years will be $(1 + i)^n$
And the total interest paid on £1 in n years will be $(1 + i)^n - 1$

3.5.3 Future value of £1 invested at regular intervals

If, rather than investing a lump sum, one invested the same sum annually, then the calculation would be different. At the end of each year the total would comprise all the previous annual investments, plus the compound interest earned on them and the next annual contribution. The formula for this is:

Future Value
= [(£annual investment + interest)$^{\text{number of years}}$ − £annual investment)]/interest

If £200 was invested annually over a period of 5 years at 4 per cent interest, then:

Future value for one pound invested
= [(£1 + 0.04)^5 − £1]/0.04 = (0.22)/0.04 = £5.50

Therefore, for £200 invested per annum the yield after 5 years would be £1100.

3.5.4 Present value of £1

The question being answered here is what would £1 in a few years time be worth today. In other words: what sum needs to be invested at the present time at a given rate of interest to accumulate £1 by the end of a given period of time. This needs to be addressed because a developer may wish to make a return of so much in n years and therefore needs to know how much has to be invested now to make this return. In effect, it is the opposite of the compound interest calculation. The formula for this is:

PV of £1 = £1/(£1 + interest)$^{\text{number of years}}$

If a return of £1000 pounds was required in 5 years' time and was discounted at 4 per cent per annum, what would have to be invested today?

PV of £1 = £1/(£1 + 0.04)^5 = £1/1.22 = £ 0.819

Therefore, the amount required to be invested today to yield £1000 in 5 years = £819.

3.5.5 Present value of £1 per annum/year's purchase

This is asking what is of a series of annual payments of £1 discounted at a given rate of interest, worth today? Put another way, the PV of the right to receive an annual income of £1 at the end of each year for a given number of years, each year's income being discounted at a given rate of compound interest. This is needed, as the developer wants to know if the total received rental annual income, discounted at a fixed interest rate over the life of the building (in perpetuity), is sufficient to make a profit after deducting all the costs of the proposed development. The formula for this is:

Year's purchase in perpetuity (YP) of £1 = 1 / interest

If the annual rental income is £1000, discounted at 4 per cent per annum, what would the gross development value be?

Year's purchase in perpetuity = £1/0.04 = 25
The gross development value would be 25 × £1000 = £25,000.

3.5.6 Present value of £1 payable at regular intervals

This would be useful for calculating the capital equivalent of regular outgoings such as maintenance, wages, rates or rents. The formula for this is:

PV of £1 payable at regular intervals
$= [(£1 + \text{interest})^{\text{number of years}} - £1] / [\text{interest}(£1 + \text{interest})^{\text{number of years}}]$

What is the present day value of £200 paid annually for 5 years, assuming an interest rate of 4 per cent?

Present day value of £1 $= [(£1 + 0.04)^5 - £1]/[0.04(£1 + 0.04)^5]$
$= [1.22 - 1]/[0.04(1 + 0.04)^5]$
$= [0.22]/[0.0488]$
$= £4.51.$

Therefore, the present day value of investing £200 per annum for 5 years $= £902$.

3.5.7 Sinking fund

It may be necessary to know how much to put aside each year to pay for the replacement of a substantial item of plant. Whilst the figures used here do not reflect the reality of such a situation, as the time scale and monies involved would be larger, the same figures have been used for comparison with the outputs of the previous formula. The formula for this is:

Sinking fund $= \text{interest} / [(£1 + \text{interest})^{\text{number of years}} - £1]$

So, if £1000 is required in 5 years, discounted at 4 per cent per annum:

Sinking fund for £1 $= 0.04/[(£1 + 0.04)^5 - £1] = 0.04/0.22 = 18.2\text{p}.$

Therefore, to achieve a yield of £1000 in 5 years it is necessary to invest £182 per annum.

References

Ashworth, A. (1999) *Cost Studies of Building*, 3rd edn, Longman.

Baum, A. and Mackmin, D. (1989) *The Income Approach to Property Valuation,* 3rd edn, Routledge

Davidson, A. (2002) *Parry's Valuation and Conversion Tables*, 12th edn, The Estates Gazette.

Ferry, D.J. and Brandon, P.S. (1999) *Cost Planning of Buildings*, 7th edn, Blackwell Science.

Jaggar, D., Ross, A., Smith, J. and Love, P. (2002) *Building Design Cost Management*, Blackwell Science

Kelly, J. and Male, S. (1992) *Value Management in Design and Construction*, E&FN Spon

Lupton, S. (2001) *RIBA Handbook of Practice Management*, RIBA.

Millington, A. (2001) *An Introduction to Property Valuation*, 5th edn, The Estates Gazette.

Ratcliffe, J. and Stubbs, M. (1996) *Urban Planning and Real Estate Development*, UCL Press.

Scarrett, D.T. (1996) *Property Valuation: The Five Methods*, E&FN Spon.

Introduction to design economics

4.1 RIBA Plan of Work

Before investigating this topic it is important to see the context in which it occurs. In the UK work is traditionally carried out to the RIBA Plan of Work, which is summarised in Table 4.1. It is often used to determine when and how much the architect and others in the design team are paid.

This document is a very succinct way of understanding the process and allows the user to focus on the relevant issues. Take, for example, environment design decisions: it is important the design team considers strategic environmental decisions such as energy strategies and the use of natural light and ventilation at the outline design stage as this will determine the orientation and footprint of the building. Generally, it is only at the detailed design stage that issues such as embodied energy of the materials used come into play when selecting alternatives.

As indicated in section 3.2.5 professional fees paid to architects, building service engineers, consultant structural engineers, quantity surveyors and others, vary from 10 per cent to 17 per cent of the final cost of the building, depending on its complexity. This figure is divided between the design team according to their contribution. Architects used to be paid as they completed various stages of the RIBA Plan of Work. For example, their proportion used to be divided as 15 per cent on completion of C, 20 per cent for D, 40 per cent for E, F, G and 25 per cent after stage H. The others involved would be paid in a similar manner depending on how much effort would be needed to comply with a particular stage of the work.

Whist many are still paid this way much has changed in the last few years, because of the different ways work is procured and the need to fast-track

Table 4.1 Adapted from the RIBA Plan of Work

Stage			Process
Briefing	A	Inception	Prepare general outline of requirements and plan future action
	B	Feasibility	Provide client with an appraisal and recommendation so as to determine the form it will take and that it is feasible functionally, technically and financially
Sketch plans	C	Outline proposal	Determine general approach to layout, design and construction to obtain authoritative approval on the outline proposals
	D	Scheme design	To complete the brief and decide upon particular proposals, including planning arrangement, appearance, constructional method, outline specification and cost, and to obtain all approvals
Working drawings	E	Detail design	To obtain final decision on every matter related to design, specification, construction and cost
	F	Production drawings	To prepare production information and make final detailed decisions to carry out the work
	G	Bills of quantities	To prepare and complete all information and arrangements for obtaining tender
	H	Tender action	To obtain main contract tenders, check and appraise them and make recommendations about acceptance
Site operations	J	Project planning	To enable the contractor to programme the work in accordance with contract conditions; brief site inspectorate; and make arrangements to commence work on site
	K	Operations on site	To follow plans through to the practical completion of the building
	L	Completion	To hand over the building to the client for occupation, remedy any defects, settle the final account, and complete all work in accordance with the project
Feedback	M	Feedback	To analyse the management, construction and performance of the project

projects, i.e. commence and overlap stages of the process before others have been completed. It is clearly in the interests of the design team not to quote a lump sum for the work unless the brief and service to be provided has been very clearly defined, which explains why there is a preference for the fee to be linked with the final cost of the project.

4.2 Factors affecting the cost of the building

4.2.1 Introduction

The following sections are written with the assumption that the decision has been made to progress the development and a thorough site survey has been carried out to reduce uncertainty. It should be noted there are often overlaps between the various sections.

Clients, especially public bodies, used to have two different cost budgets for buildings, namely the capital cost of the building and the maintenance and running costs. These were almost invariably kept separate with the result there was no incentive to design buildings that were efficient during their lifespan. This has gradually changed so that now emphasis is placed on making design decisions to take account of these issues. More detailed analysis is given in Chapter 6.

4.2.2 The site

No two sites are the same and some of the issues have already been raised in section 3.2.4.

- The shape and area of the site is the significant determining factor as this sets the parameters as to the possible footprint in terms of shape and orientation, and the number of storeys that will be required. Restricted sites also affect the amount and way materials can be stored during the construction process and limits the type and positioning of accommodation for site personnel. It may be necessary to find alternative accommodation close to the site.
- The topography of the site is important as sloping sites generally mean more expensive building.
- Poor ground conditions often will mean more expensive foundations and temporary works may be required during construction either to support the ground during excavation or to keep water at bay if the water table is high.

- Other influences on substructure costs include the cost of dealing with contaminated ground, hidden services, underground structures and archaeological finds.
- The location of the site can have a significant effect on the construction costs. In a congested urban area there may be restrictions on the hours of working, the times vehicles can offload, difficulties in access, levels of noise permitted and trespass problems associated with tower cranes passing over neighbours' properties (see *Operations Management for Construction*, section 1.1). Rights of way, such as pavements, may have to be temporarily diverted or covered to protect the general public from falling objects. Sites situated in rural surroundings may incur extra transport costs and access routes may be too weak to accept the heavy transport without some widening or strengthening.
- Surrounding buildings may have to be protected from damage from the construction work.
- Any existing buildings on site will have to be demolished before new building can begin.

4.2.3 Size and scale of the project

Whilst sections 4.2.4 to 4.2.7 can distort the cost implications of this section, generally speaking, the larger the project for a particular use, the cheaper the unit cost becomes. The reasoning behind this includes:

- Larger projects can be more efficiently managed and completed faster as it becomes more economical to use sophisticated plant and equipment. Sub-contractors put in more competitive bids because of the size of their contract, and materials can be obtained cheaper due to quantity.
- As the size of a project diminishes, the design cost as a proportion of the overall cost increases as the time taken for the design is disproportionately higher.
- Providing the overall unsupported floor span does not become excessive, the larger the floor plan area, the cheaper the building becomes. This is because the costs of providing external walls and, if load-bearing, their foundations, is relatively expensive. This is referred to as the wall-to-floor ratio.

4.2.4 Usable and non-usable space

The term non-usable space is used to describe areas of the building that are primarily non-productive. This usually means circulation spaces, which include corridors, lifts, stairwells, escalators and floor areas within rooms/ spaces needed for the passage of people and goods. The developer will obtain the greatest financial return from the building if this non-usable space can be kept to a minimum. Other issues are:

- Often open-plan design offers a lower ratio of non-usable space.
- With the exception of very small buildings, such as a private house, some protected circulation space will be mandatory to permit safe evacuation from the building in the event of fire.
- Circulation spaces used in the event of fire will have to be of a minimum size to comply with fire regulations.
- A case can be made in some buildings to have a larger than required circulation space, especially in entrances and lobbies, so as to portray a grand image to visitors and users.
- Complicated and irregular footprints usually require a greater proportion of non-usable space. This is why, for example, hotels tend to use simple, long rectangular accommodation blocks with a corridor running down the centre, with the bedrooms opening on either side, thereby reducing the proportion of non-usable space.
- In complex situations space planners (see section 1.8.2) are brought in to assist in optimising the remaining usable space.

4.2.5 Plan of building or footprint

This is a very complex problem. Much of the footprint may already be determined either by the site itself, such as a confined city site, or by the very nature of the business, such as a supermarket or car-manufacturing plant. Environmental issues may also come into play. If the developer requires an energy-efficient building using predominantly natural light and ventilation, then the distance from the external wall to the back of the room is limited to approximately 6 to 7 metres. This means if using a central corridor for circulation, the maximum distance from the two external walls is approximately 15 metres. Those working in the space furthest from the windows may need some artificial lighting depending on the nature of their work. If the footprint of the building is in excess of this then atriums could be considered as an alternative to providing artificial lighting and ventilation.

Generally, because of the wall-to-floor ratio, a square or rectangular building is the most economical to build as any change of direction complicates the roof details and sometimes the foundations. A complex footprint can slow down the production process of cladding – especially if constructed of brick and block – because of the number of changes of direction. Square buildings are not an economical solution if the site has a significant slope and, in this case, it is generally accepted that rectangular buildings following the direction of the slope are more economical.

If the development comprises more than one building, there are economies if the buildings can be linked in some way as this reduces the number of external walls and foundations that have to be constructed. For example, it has been estimated that for the same comparable area, a semi-detached house is 6 per cent more expensive than a terraced house.

4.2.6 Height

Generally, the taller the building, the more expensive it is compared with low-rise construction. However, it may be necessary to build high because of limited site area, scarcity of land or and costs are so high they dominate the cost equation making it more economical to reach for the skies. This is often the case in major world cities. The key reasons for this extra cost are:

- The structure and the foundations have to be more substantial than low-rise. This is because the structure has to support the extra height and resist the horizontal wind forces.
- To overcome the problem of disturbing the structural equilibrium of adjacent existing buildings it may be preferable to create basements. This creates extra floor space that may be of benefit to the developer.
- Circulation spaces have to be increased in area to evacuate the greater number of building users and to satisfy fire regulations.
- Lifts, with their inherent requirements, will have to be provided.
- The building will be more highly serviced using more sophisticated and larger equipment, which in turn has to be installed by specialist contractors.
- The construction costs are higher because of the need to provide cranes and hoists to lift and transport materials and labour.
- Long-term maintenance is likely to be more expensive because of the extra risk of deterioration at high levels, difficulty of access to carry out the works, and more complex equipment installed for the building services.

In spite of these reasons there are occasions where buildings a few storeys high offering the same usable area may be more economical than a single-storey building as some costs go down or remain constant as the footprint diminishes and the numbers of storeys increases. Examples of these are as follows:

- The roof covering diminishes significantly, though not completely in proportion, as circulation spaces increase. For example, a two-storey building only requires a roof area of about 50 per cent of the single-storey building of the same usable total floor area.
- The increase of the foundation design may not be substantial until the weight of the multi-storey solution requires a different design solution.
- Other than for disabled access and the movement of heavy or bulky goods, lifts will not be required.
- Internal finishes remain similar irrespective of height.

4.2.7 Storey height

The floor-to-floor dimension is determined by several factors, such as minimum heights laid down in the building regulations, the ability to accommodate tall pieces of equipment and plant required for the user's business, space to cater for building services such as ventilation ducts, and raised floors for information technology (IT) services. Other buildings such as sports centres, swimming pools, tiered lecture theatres and conference facilities will require extra height because of their specific design requirements. Any increase in storey height adds to the overall cost of the building as it increases the quantities of materials used and adds to the weight of the building, which in turn has to be supported by the structure and foundation. There are also energy implications in running the building, as there is a greater volume of air to be heated or cooled, not to mention the thermal capacity implications depending upon the materials selected for the construction of the building.

4.2.8 Buildability and construction

This is in part linked to value engineering (see Chapter 7), but in essence is to do with how the design has taken into account the problems of construction and produced solutions to assist the construction process, thereby making it more productive and more economical. Some examples are:

- To design so standard-sized components can be used without adaptation on site, or alternatively agree with the manufacturer to produce the required size.
- To have as much repetition as possible, rather than many different sizes, without compromising aesthetic considerations.
- Consider the sequence of operations in terms of productivity and continuity of work.

- Ask the question 'In practice, how is what I want produced?' and if the answer proves to be 'With difficulty', then ask 'Can I modify my design accordingly?'
- Is it possible to modify the design and methods of fixing so all exterior work can be carried out from within the building so no scaffolding is required on the outside of the building?
- If possible, bring suppliers', sub-contractors' and contractors' production expertise in as early as is practical into the design process.
- Consider the use of off-site production. Sometimes the components produced can be more expensive, but savings can be made in the increase in the speed of construction that results.
- Is there a freely available source of both material and skilled labour of the kind required to complete the building?

4.2.9 Deconstruction

Increasingly, consideration is being given to ease of dismantling the building so as much of the material can be recycled when the building comes to the end of its useful life. There are some serious implications for the designer in this concept. Not only must the designer consider how components and structure are taken apart, but many of the newer construction techniques designed to combat skill deficiencies and improve productivity make segregation of materials more difficult, such as gluing architraves and skirting. Equally, many of the completed multi-material components are not readily dismantled. At first glance, this could mean increased construction costs unless alternative design solutions are provided, but this has to be balanced against the longer-term issues. It may be that as designers consider facilities management issues in more detail, the need to replace, repair, maintain or alter the building, may also have a positive knock-on effect on the economics of deconstruction and recycling.

References

Ashworth, A. (1999) *Cost Studies of Building*, 3rd edn, Longman.
Lupton, S. (2001) *RIBA Handbook of Practice Management*, RIBA.

Approximate estimating

5.1 Introduction

The techniques of approximate estimating can be applied for several reasons in the development of the project. First, as a feasibility study to determine whether or not the project is financially viable. Second, for budgeting the project as a means of establishing the financial implications throughout the project from design to completion, so that the client is aware of the financial requirements, known as the cost plan, and also to assist in obtaining funds to finance the project. Third, for estimating in providing information to assist management in deciding whether or not their company wishes to tender (see section 10.4).

5.2 Methods

Four methods are available, the accuracy of each depends on the accuracy and quantity of information and data available. These are:

- functional unit method or unit of accommodation method
- superficial (floor area) method
- elemental cost plan
- approximate quantities.

Table 5.1   Functional unit costs

Building type	Costs per m^2	Building type	Costs per m^2
General hospitals	£630–1330	Middle schools	£450–920
Swimming pools	£850–1800	Student residences	£500–1200
Public houses	£520–1300	Hotels	£250–1000

Functional cost	Area m^2	Cost £1000s
District general hospital	65–85/bed	63–94/bed
Theatres		7.3–10.5/seat
Secondary school	6–10/child	4.2–8.35/child
Large student residences (200 plus rooms)	18–20/bedroom	10.5–19/bedroom

5.3 Functional unit method

This method is largely used in the public sector where significant sums of money are involved, but very little (if any) design work has been executed.

Examples of a functional unit are the number of hospital beds, student residential accommodation places, and high- or low-risk prisoners, hence also the term 'accommodation method'. Having established the number of places required, this number is then multiplied by a figure based on the cost of providing such a place (see Table 5.1). This cost is calculated from previous provision and adjusted, making allowance for inflation, changes in specification and design, the market forces at the time, land costs, etc. It can rarely be accurate as a method because of the many intangibles, not least the condition of the land it is being built on and availability of services and infrastructure.

5.4 Superficial area method

This is an improvement on the previous method, but relies on some elementary design work being carried out. This is a method readily understood by developers, designers and builders alike.

It is necessary to know the superficial area of the building (the footprint), the number of storeys and any significant differences on any floor level. It also requires the type of usage for the areas involved if they are different. No deductions are made for stairwells, lift or circulation spaces. This superficial area in square metres is multiplied by an approximate figure based upon the usage of the building. This will vary depending, for example, on the amount of services or quality of finish that the building type may require. As can be seen in Table 5.2, the range for each building type is very large, so further

Table 5.2 Superficial area costs

Building type	Costs per m^2	Building type	Costs per m^2
Hospitals	£700–1400	Schools	£400–1000
Swimming pools	£900–2000	Student residences	£600–1400
Public houses	£500–1400	Hotels	£300–1000

adjustments to this multiplying factor will be made to take into account of such variants as location, quality of specification, complexity of shape, number of storeys, and suspected ground conditions including substrate and topography.

5.5 Elemental cost plan

This can be produced from the designer's preliminary drawings. To do this, a list of the elements of the building is drawn up and an elemental cost added. This latter information comes from measuring the size/amount of the element and multiplying by a unit rate. This rate will be based, in the case of the contractor on the experience of the company and estimating staff employed, and for the quantity surveying practice on its previous expertise and standard works, such as *Spon's*, *Laxton's* and the BCIS *Standard Form of Cost Analysis* or the BCIS online (www.bcis.co.uk).

Alternatively, rates used on similar buildings and proportioning accordingly as shown in Table 5.3, which is adapted by permission of the Chartered Institute of Building, from the *Code of Estimating Practice*, endorsed by the CIOB. In this case Building A has already been built and its overall area was 2,900m². The proposed building, B, is 3850m². Prices can be adjusted to take account of obvious variances between the two buildings. For example, in this case in the Substructure*, building A used reinforced concrete pad foundations and the case of building B, piled foundations.

5.6 Approximate quantities techniques

Accurate quantities are preferred, but this may not always be possible, as the design may not have reached a sufficient development to provide this detailed information. In such cases, approximate quantities can be used. Examples of where they may be used are where speed is the key issue and all the detailed production drawings cannot be produced in time and where the ground surveys have not been completed, or there are other unknown factors. If it is decided to use approximate quantities, there is a contractual provision using

Table 5.3   Elemental cost plan

Element	Elemental costs £/m²	Building A	Elemental costs £/m²	Building B
Substructure*	68.10	197,490	95	365,750
Steel frame	68.81	199,549	67	257,950
Upper floors	36.21	105,009	36	138,600
Roof	50.43	146,247	50	192,500
Stairs	14.22	41,238	14	53,900
External walls	62.50	181,250	63	242,550
Windows and external doors	46.55	134,995	47	180,950
Internal walls and partitions	31.03	89,987	31	11950
Internal doors	6.59	19,111	7	26,950
Wall finishes	13.36	38,744	13	50,050
Floor finishes	61.21	177,509	61	234,850
Ceiling finishes	17.67	51,243	18	69,300
Steelwork (fireproofing)	3.88	11,252	4	15,400
Fittings and furniture	9.48	27,492	9	34,650
Sanitary appliances	5.60	16,240	6	23,100
Disposal installations	4.40	12760	4	15,400
Water installations	6.03	17,487	6	23,100
Heat source and space heating	52.16	15,1264	52	200,200
Ventilation and cooling	20.69	60,001	21	80,850
Electrical installation	103.45	300,005	103	396,500
Lift installation	0	0	13	50,050
Security alarms	27.28	79,112	27	103,950
Fire alarms	10.73	31,117	11	42,350
Builders' work in connection	7.03	20,387	7	26,950
Minor works	3.10	8,990	3	11,550
Total elemental cost	728.53	2,112,737	768	2,956,800
Preliminaries for building	58.31	169,099	58	223,300
Net building cost	786.84	2,281,836	768	2,956,800
Site works	105.39	305,631	105	404,250
Drainage	24.18	70,122	24	92,400

Table 5.3   continued

Element	Elemental costs	Building A	Elemental costs	Building B
	£/m²		£/m²	
External services	7.16	20,764	7	26,950
Preliminaries for external works	8.21	23,809	8	30,800
Net scheme cost	931.78	2,702,162	912	3,511,200
Design fees	44.83	130,007	45	173,250
Statutory fees	11.03	31,987	11	42,350
Overheads and profit	55.17	159,993	55	211,750
Contingencies	34.47	99,963	35	134,750
Budget total	1077.28	3,124,112	912	3,511,200

the Standard Form of Contract With Approximate Quantities prepared by the Joint Contracts Council (JCT).

There are three different ways of achieving a detailed measure of the actual quantities used in the construction:

- the work is measured at the end of the contract;
- the work is measured as soon as the production drawings are completed and then substitution bills are produced; or
- the work is measured as it is constructed.

Approximate quantities give a more accurate estimate of the likely cost of the building than the previously described methods, but can only be used if there is sufficient detailed design completed to carry out the calculations. Similar in approach to the production of the bills of quantities, the sequence of abstracting the information is usually the same, as far as it can be, to the Standard Method of Measurement 7 (see Chapter 9), but in this case the unit of measure is a composite of several smaller units and only measures the main or significant elements of work to be constructed.

Two examples of what is meant by a composite rate, derived with permission from *Laxton's Building Price Book*. Brick manholes, described as:

Excavation, 150mm concrete bed, 215 class B engineering brick walls, 100mm clayware main channel, concrete benching, step irons and cast iron manhole cover.

The costs being a function of size and depth such as

600 × 450, two branches, 750 deep	number	£513
750 × 450, three branches, 1500 deep	number	£866
900 × 600, five branches, 1500 deep	number	£1245

Strip foundation described as:

Excavating, 225mm concrete, brickwork, dpc, common bricks.

800 deep for 103 wall	metre run	£54
800 deep for 215 wall	metre run	£104
1200 deep for 215 wall	metre run	£142

References

Aqua Group (1999) *Tenders and Contracting for Building*, 3rd edn, Blackwell Science.

Brook, M. (1998) *Estimating and Tendering for Construction Work*, 2nd edn, Butterworth Heinemann.

BCIS (2008) *Standard Form of Cost Analysis*, 3rd edn, BCIS.

CIOB (1997) *Code of Estimating Practice*, 6th edn, Blackwell.

Laxton's Building Price Book, Butterworth-Heinemann.

Smith, A.J. (1995) *Estimating, Tendering and Bidding for Construction Work*, Macmillan Press.

Smith, R.C. (1999) *Estimating and Tendering for Building Work*, Longman.

Spon's Architects' and Builders' Price Book, E&FN Spon

Whole-life costing and life-cycle assessment

6.1 Definitions

There can be confusion in the way the terms life-cycle assessment (or analysis) and whole-life costing are used. The author uses the following definitions for the purpose of this text.

> Whole-life cost (WLC) is a tool to assist in assessing the cost performance of construction work, aimed at facilitating choices where there are alternative means of achieving the client's objectives and where those alternatives differ, not only in their [initial] costs but in their subsequent operational costs (from the ISO Standard 15686-5, Buildings and constructed assets – Service life planning).

> Life-cycle assessment (LCA), sometimes referred to as life-cycle analysis is, 'a method to measure and evaluate the environmental burdens associated with a product system or activity, by describing and assessing the energy and materials used and released to the environment over the life cycle' (Edwards *et al*. 2000).

6.2 Introduction

Traditionally, buildings were looked at financially in two distinct ways: capital costs and maintenance costs. The two were rarely linked. The effect of this was that many buildings, especially in the public sector, were built as cheaply as possible within the budget allocated for this purpose, often using inferior materials and components which needed replacing at frequent intervals thereby raising the maintenance budget.

The industry has now moved to looking at the costs incurred throughout the life of the building as it is argued that the client, although initially needing to find more capital money, would, in the long term, pay out less for the building over its lifetime as a result of lower maintenance and running costs. For example, by increasing the insulation and using low-energy light bulbs, the capital costs of proving this combined with the energy consumption during the life of the building would be less than building to minimum insulation standards and using normal bulbs. There would also be a further saving, as the boiler used to drive the central heating would need to have a smaller capacity.

The problem with WLC calculations is the number of variables that have to be considered inevitably make the outcome figures suspect, especially when predicting costs over a long timescale. For example, who can predict what interest rates are going to be 15 years from now? However, it is important at the design stage to make environmental decisions that impact on the outline and detailed design of the building. Designers and developers must consider such issues as the orientation and footprint of the building, the energy strategy, and selection of materials with an eye on both the impact of the building itself and the impact resulting from running and maintaining the building. The developer will want to know if the sustainability decisions stack up financially; they probably will.

This subject area has become more relevant with the increasing trend towards developments, such as Private Finance Initiatives (PFI), which require that the building is constructed and maintained by the contracting organisation for 25 years or more – known as facilities management.

It is also important to relate these decisions with value engineering (Chapter 7) which can be used as a tool for WLC and LCA. This is because value engineering applied to a building at the design stage will have possible implications on the long-term maintenance and use of the building.

The first part of this chapter considers the cost effectiveness of a building in terms of its capital cost and maintenance and running costs, i.e. whole-life costing, but it should also be remembered that cost effectiveness to the user/owner includes other issues besides the cost of the building. For example, an airport is also about the number of long-term parking spaces available, ease of setting down and picking passengers up, the speed at which passengers can be processed, speed of baggage handling, security checks and processing and so on. In other words the cheapest building is not always the most effective in use.

6.3 Cost centres

In a whole-life costing exercise it is necessary to consider all the expenditure that may be required throughout the life of the building. All of these are affected by the design decisions. The main categories of costs (RICS 1986) are:

- capital costs
- financing costs
- Operation costs
- annual maintenance costs
- intermittent maintenance, replacement and alterations costs
- occupancy costs
- residual values and disposal costs.

An alternative approach to categorising (Kelly and Male, 1993) is:

- investment costs
- energy costs
- non-energy operation and maintenance costs
- replacement of components
- residual or terminal costs.

6.4 Period of analysis

The life expectancy of a building in terms of its physical state and its usefulness needs to be considered by the designers. After World War II, prefabricated houses were produced with only a 5- to 10-year life expectancy. Some of these are still in use and indeed a few have become listed buildings. In the early 1960s many clients believed buildings should only be designed with a 20-year life expectancy after which they would be demolished. In Hong Kong in the property boom of the 1970s, the rule of thumb was that when the cost of the land the building was built on rose to equate to the cost of the building, it was torn down and another, usually taller, building was constructed. On the other hand the cathedrals built over history, including the more recent examples, such those at Coventry, Guildford and Liverpool, have been designed with a very long life expectancy.

Another approach is to consider building with a more flexible design so that changes in use can occur. For example, the large Georgian terraced houses have subsequently been converted into uses other than housing, such as offices, hotels, nursing homes, small schools and hospitals. Whilst there remains an attitude to design and build for the initial purpose intended, some developers are thinking ahead about revised usage. The Trafford Centre in Manchester is an example of this.

The period of analysis is also a function of how long the investor is interested in the building. For example, a facilities management contract on a PFI contract may be 25 or 35 years. The developer/investor may not be concerned about what happens after that, although the final owner, e.g. the government, may have a different view on this. The developer will also be influenced by whether or not they are developing with a view to sell, rent or use themselves.

6.5 Factors affecting life expectancy

6.5.1 Physical deterioration

The physical state of any building will decline over the passage of time. The fewer preventative measures taken, the quicker this will take place as can be seen in the rapid state of deterioration once a building is left unoccupied and nature takes its course. However, if a building is well maintained, it will normally last until it is demolished. The rate at which buildings deteriorate depends on many factors, including:

- the developer setting inappropriate standards of performance specification
- a lack of appreciation of the causes of degradation
- poor design detailing
- inappropriate selection of materials and components
- poor construction practice
- inadequate maintenance schedules
- damage and vandalism
- a different use by the owners and occupiers from the original design.

6.5.2 Economic obsolescence

This is in part related to some of the following causes of obsolescence (sections 6.5.3–6.5.7) as economic analysis will be a determining factor in accepting the need to vacate or demolish the building.

The example of Hong Kong mentioned above is an example of economic obsolescence. Here the potential value of the land and new development is worth more than the existing rental income from the existing building. In other words more profit can be obtained from the site by demolishing and building something else which offers a greater return.

6.5.3 Functional obsolescence

Buildings are usually designed with a specific purpose in mind. If that use changes or ends then the building may not be appropriate for another use. The Millennium Dome is a classic example of how difficult it can be, remaining empty for many years. It is not always the case however. The city of Manchester, among others, is converting many Victorian industrial buildings into retail stores and residential accommodation to bring life back to the city by attracting tens of thousands of tenants or owner-occupiers. If no alternative use can be found, then the building will eventually be demolished.

6.5.4 Technical obsolescence

As with most items produced, technology moves on and the product becomes technically dated. Buildings are no different. Some components wear out faster than others and have to be replaced. New technologies for components become available which make the building more efficient and easier to maintain. For example, single-glazed untreated softwood timber window frames are commonly replaced by double-glazed uPVC frames.

The building will usually have been designed for a specific purpose, but as the user's technological requirements alter as a result of the latest plant and equipment, different handling requirements and the development of information technology systems, the building may be ill equipped to cope with these changes.

6.5.5 Environmental obsolescence

If the building is old the insulation standards may be such that the cost of heating the building has become prohibitive, especially if the processes now being used do not emit the same levels of heat as previously. This, coupled with other factors, could make the building obsolete for the owner.

6.5.6 Social obsolescence

Some buildings such as many of the multi-storey blocks built in the 1960s to resolve the post-war housing problems became socially unacceptable and many have now been demolished and replaced with low-rise housing. Buildings cannot readily be moved from one location to another and many churches and chapels have also become redundant, either because of falling congregations or changing populations. For example, many immigrant populations of the Muslim faith have gradually filled inner town and city areas previously predominately populated by Christians who have moved out into the suburbs. Some of these buildings are now being used as mosques, but many have been sold on and converted for other uses other than for worship.

6.5.7 Legal obsolescence

This occurs when the building is unable to meet the current regulations without substantial costs being incurred which outweigh the benefits to the building owner. This could be a structural defect making the building unsafe, the discovery of substantial amounts of damaged asbestos which have to be removed or sealed in, not satisfying the current fire regulations, and more recently, legislation concerning disabled access.

6.5.8 Aesthetic and visual obsolescence

Public tastes changes over a period of time. What was once acceptable now becomes unacceptable and sometimes vice versa. Some seem to transcend time and are always liked. The 1960s was a period in UK history where there was an explosion of ideas and experiments right across the arts, and architecture was no exception to this. Unfortunately, whereas in the other art forms mistakes could be hidden or destroyed and then forgotten about, disliked buildings remain for a much longer time because of economic considerations. However, often in the end public pressure can have an influence on the decision to demolish the building.

6.6 Data for whole-life costing

6.6.1 Historical data

Those concerned with the running and maintaining a building collect data over a number of years. These can include alterations, replacement, general maintenance, energy consumption, and so on. If the data are collected conscientiously, then this can be very useful source of information. The problem is that the data may have been collected for different reasons. For example, service engineers may be collating an energy audit, cleaning may come under a different section, and alterations and building maintenance comes under small works. The way costs are presented may also be unhelpful. It is more useful, when calculating the WLC, to know the cost of painting a square metre of wall rather than the total cost of decorating an executive suite (for instance), so that this data can be used for other applications. It is important that from the point in time it has been decided data should be collected for future WLC purposes, the method of collecting is standardised so it is user friendly.

6.6.2 Experience

Never underestimate the value of experience. It is not always quantifiable, but it is not unusual for the experience of a key person in a section who has been there for a long time to often reliably predict the life of a component, piece of equipment, and in particular how long operations may take to complete. This experience was tapped into in 1992, when a number of building surveyors were asked to fill in a questionnaire estimating, using their experience, the life expectancy of a variety of components. The results were published as RICS (1992) *Life Expectancies of Building Components*.

6.6.3 Manufacturers' literature

Manufacturers will often offer guarantees as to the life of their product. If they are a reliable and well-established company, they will not offer guarantees of life expectancy above that which they, from their experience, can offer as this could result in court action against them with excessive damages being awarded. Their guarantee can therefore be taken as a minimum life before any remedial or replacement work will be needed. How much is added to this relies to a certain extent on knowledge gained from experience and reading trade literature and magazines. Using guarantees from small organisations is dangerous as the owner may be out of business before the guarantee expires.

6.6.4 Research databases

Increasingly, informed databases are being produced by various organisations such as the Building Research Establishment, but these are only accessed if paid for. However, the joining costs are relatively small compared to the savings to be made on a significant project.

6.6.5 Calculations

An example of the use of calculations is the annual amount of energy needed to heat/cool and light a building once the basic footprint and construction of the exterior elements has been decided. This calculation would also have to take into account the usage of the building in terms of days per week or per year, the costs of reheating after cooling down over a weekend/holiday, likely solar gain, the type of work being carried out in the building, and the numbers of people using it. This would not take account of plant efficiency, selection of fuel, changing fuel prices and so on, but serves as a useful guide as to the likely costs.

6.7 Issues to be considered at the design stage for whole-life costing

6.7.1 Maintenance

This is to do with maintaining the physical integrity of the materials, components, plant and equipment and should not to be confused with operating costs such as cleaning (section 6.7.3). It can be described in two categories:

- *Planned maintenance.* This is when the maintenance is carried out on a pre-determined time scale, similar to the maintenance logbook provided with a new car. It may be for aesthetic and comfort reasons such as decoration, to ensure the highest possible efficiency from plant, or to protect the basic fabric from deterioration and eventual failure. A good example of the latter is replacing flat roof coverings before the roof leaks.

- *'Crisis' maintenance*. This is dealing with the unexpected, such as a burst pipe, or the expected but unpredictable, such as light bulb failure. The term is sometimes applied to maintenance protocol that has no planned maintenance but just reacts as needed.

Planned maintenance can be costed relatively easily using today's prices. It is then necessary to apply the appropriate formulas to estimate the future costs in real terms (section 3.5) and then it can be established what finance is required to pay for it. Crisis maintenance can be estimated, but relies much more on experience and historical data for buildings of a similar nature. Whether or not crisis maintenance is cheaper in the long run than planned is a different debate, but the advantage of planning is the disruption to building users is likely to be included in the thinking processes. It is important the designer takes maintenance into account, especially ensuring the replacement of components can be easily and rapidly executed so as to reduce costs and disturbance.

The costs of maintenance vary depending upon the use of the building. Buildings which are highly serviced, such as hospitals, will need more maintenance then those with less. It has been suggested this can vary from between 5 and 30 per cent of the occupancy costs.

6.7.2 Energy

There are several factors which affect the energy costs of the building:

- The initial capital costs of the plant and equipment that consume energy such as for heating, cooling, ventilation, de-humidification, lighting, lifts, escalators and kitchen equipment.
- The efficiency of the plant. Some plant is more energy efficient than others.
- The staff required to operate the plant and equipment, such as boiler men. These will have to be employed throughout the life of the building or until replaced with automated plant.
- The staff required to maintain the equipment. This service could be outsourced, especially with the developing expertise in facilities management companies. Much will depend upon the frequency of maintenance and the skill required to carry out the work as against the convenience of having in-house staff.
- The cost of parts and replacement equipment. The design decision is whether or not it is worth while spending more on a piece of gear now and replacing it less frequently than selecting a cheaper one which

requires more frequent changes. It should also be noted that equipment technology continues to make advances, so changing certain plant at more frequent intervals might offer significant energy savings, which would compensate for the replacement costs.

- The cost of providing access to maintain/replace the plant can be a significant issue. If the plant requires a large amount of space for this to occur, but the event only occurs once every 20 years, it might be cheaper not to provide this costly space, which will be redundant throughout this period. Users/owners will then have to accept the disruption and extra cost of perhaps having to remove elements of the building to carry out the work.
- Selection of fuel. This is a very difficult issue. For example, in the late 1960s, the tendency was to use oil for heating, as it was readily available and very cheap. Then in the early 1970s, when OPEC was formed, oil prices rocketed and alternatives were sought, notably natural gas. With the deregulation of the energy suppliers and the ability to source energy from different suppliers, it has become more difficult to estimate what costs will be in the future.
- The footprint and orientation of the building is extremely important. The former as it will determine the amount of natural ventilation and lighting the occupiers can use as against requiring artificial lighting and mechanical ventilation; and the latter in terms of the effects, or otherwise, of solar gain and wind chilling.
- The location and siting of the building exposes the building to different climatic conditions. Windy sites will have a cooling effect on the building whereas well-protected sites may require mechanical ventilation and air-conditioning especially in the summer. Temperature variations will also occur depending upon the height above sea level and geographical location.
- Energy consumption is a factor of many of the above, plus the design of the outer fabric. This is primarily about the amount of insulation provided, elimination of cold bridging, and control and utilisation of solar gain. It is also a function of the use of the building. For example, the greater the number of people and amount equipment in the building, the more heat generated with potential overheating problems.

The energy costs of a building are on the order of 20 to 35 per cent of the occupancy costs, but there is no reason why this figure could not be reduced below this with sensible design decisions and innovatory solutions.

6.7.3 Cleaning

Cleaning is an important issue for the general welfare of the staff and the company's image. Few like working in untidy, dirty conditions and an unkempt working environment gives off the wrong signals to potential and existing clients. The design process largely determines the cost and method of cleaning.

A significant issue, often neglected, is that the selection of a particular material at the design stage determines the method of cleaning, which in turn may determine the solutions used for cleaning, some of which have potential hazards associated with them (see *Operations Management for Construction*, section 4.9.3).

The selection of materials or components also affects the frequency and ease of cleaning. Can the windows be cleaned from inside or are ladders, cradles, etc. required to carry out the work from the outside? How easily does dirt become engrained in a surface and can the surface be washed, dried and polished in one operation?

A final consideration is whether in the future, the use of the space will change either by function or numbers of occupants (see also 6.7.4). The costs of cleaning are between 10 and 20 per cent of the occupancy costs.

6.7.4 Communications

Telecommunications provision has revolutionised the way building users function, especially in commerce, industry, education and health and is set to continue. It is suggested that not only will the equipment become more sophisticated, but it will also change the way we work with increasingly more people working from home for at least part of the week. All this means that systems will be replaced at regular intervals with more up-to-date equipment and the way office space is used will also change.

There is the possibility that as a result of home working, more 'drop-in' office space will be needed, where employees do not have their own personal space, but just find an available desk. In turn this means storage of personal items and files, etc. will have to be considered.

Currently the use of raised floors in open-plan offices permits minor alterations to office layouts to occur relatively easily provided fire exit routes are incorporated or maintained. However, if substantial changes are required as a result of the previous comments, the costs of modification would increase.

6.7.5 Alterations and flexibility

Following on from communications, it can be expected in the life of most buildings, except residential, there will be a lot of alterations. The more potential for flexibility that is built in, the less cost will be incurred later and the greater the opportunity to change the use of the building for new tenants/users of the building.

6.7.5 Security

This is an uncertain area as the need for security systems has only become apparent in recent years. A key issue here is to allow for systems to be updated as and when required without disrupting the building users. However, it should be noted that especially since September 11, 2001 changes to layouts and means of access will also have to be considered. This will inevitably be an extra financial burden.

6.7.6 Financial advice – rates and taxes

This is the province of specialist advisors. Tax law changes each year as a result of the Chancellor of the Exchequer's budget. It can affect areas as diverse as capital gains, maintenance depreciation and incentives for employment in the area the building might be located.

6.8 Is whole-life costing effective?

If capital costs and maintenance are separated because the developer is selling on, then it is likely that as the developer is only concerned with the profit made on the capital cost of the building, WLC issues will not be at the top of the agenda. On the other hand, if the contract includes both construction and facilities management the attitude will be different.

WLC has it limitations because of the high number of unpredictable variables used in the equations especially that of anticipating inflation and interest rates. Long-term investments are also at the vagaries of the stock market. Since the life of the building is expected to be anything from 30 to 100 years this makes cost predictions difficult. Therefore if negotiating a facilities management contract it is necessary to have the opportunity to renegotiate new rates every few years.

It has been argued in some quarters that volatile inflation is a problem, but this argument is suspect as interest rates tend to track those of inflation, the difference between the two remaining relatively constant. For example, if inflation is running at 15 per cent, interest rates might be 18 per cent, but as inflation falls to 5 per cent, interest rates drop to 8 per cent. Hence in real terms there is little difference. However, irrespective of this, not all building costs increase at the same rate as inflation. This is because products may come from parts of the world with different rates of inflation. Some building work, such as maintenance, which is labour intensive, may be carrying a higher rate of inflation than where there is a high material content.

It is difficult to predict the future. For example, poor design detailing and workmanship will demonstrate failure earlier than anticipated. The projected planned maintenance schedule is based upon the assumption the building owners will keep to it, but they may not, initially saving money but potentially building up a backlog of work which may be more expensive later.

However, one can get tied up with the academic rigours of precise and accurate calculations at the expense of raw common sense. Unless, after calculation, differences between two solutions are marginal, the fact one solution is clearly an improvement on the other is surely all that matters. Whilst the figures produced may be incorrect in reality, if there is a clear indication which of the two (or more) solutions is the more cost effective, then the selection becomes obvious assuming all the variables have been considered. Using different likely variables and seeing how the sums work out can be tested mathematically. It is an inexact science and there is much debate as to its validity of the calculations and readers must draw their own conclusions. This naturally leads into including the environmental issues. This is referred to as life-cycle assessment (LCA). It is argued that following this route rather than the economic debate is more fruitful especially when investigating long-term sustainability issues.

6.9 Life-cycle assessment

LCA is an analytical tool that attempts to assess the material content and the majority of environmental impacts of any manufactured item. In construction it can be used to assess materials, sub-components, components and the whole building. To understand the full impacts of a particular component or material to be used in a building requires considerable research and is normally outside the time constraints of routine building design and material selection.

There have been calls for manufacturers to publish LCA data on their products, but this is not required of a manufacturer at this time. There is also no international agreement over LCA techniques, which makes comparisons difficult especially since many products and components used in construction are imported and used directly or incorporated into other products, or exported and then re-imported after being processed elsewhere. There are various LCA systems in use and under development such as the BRE material analysis computer-based tool, ENVEST 2 and the BEES programme in the USA. These are making significant advances and simplify a complex analysis and provide an output that is specific to the project in hand rather than the generic information that otherwise is all that is available.

However, in spite of these developments there are still a number of questions that need to be asked.

- What level of detail should be included? Are nails and screws included, for example?
- The impact various materials have over time in relationship to the whole building. For example, the building's life is 60 years, the roof tiles are 30 years and the boiler is 10 years.
- The relative weighting of the various impacts. How seriously is global impact such as ozone depletion compared with a local impact such as river pollution or ground contamination?

To give a further indication of the problem, Table 6.1 considers the various impacts materials can have.

6.10 Whole building environmental assessment

Life-cycle assessment is aimed at materials and components whereas whole building environmental assessment is concerned, as the name implies, with the whole building. LCA is more comprehensive if conducted thoroughly, but over the last decade methods of assessing the overall environmental performance of a single building have been developed, notably BREEAM. These are attempts to formalise environmental impact analysis of a building, usually to ensure conformity of approach in order to grant a certificate of label. This is necessary where claims of 'greenness' need to be substantiated and compared. These types of tools usually attempt to assess environmental performance in terms of the environmental issues such as demonstrated in Table 6.2 which along with Table 6.1 are extracted from Curwell *et al.* (2002)

This is often achieved via indirect factors, e.g. the operational energy consumption of the building as an indicator of CO_2 emissions. In terms of building materials and components, the early systems usually consisted of a

Table 6.1 Key factors effecting environmental assessment of building materials

Process	Issues
Upstream – extraction and manufacturing	Energy involved in extraction and manufacturing processes
	Transportation from source to manufacturing plant
	Depletion of resources: how much reserve remains
	The amount of despoliation caused by the manufacturing process
	Quantity of waste generated by extraction and manufacturing
	Quantity of pollutants generated during these processes
	Proportion of the product made from recycled materials
Construction	Energy involved in the construction process
	The distance the material has to be transported to site
	How much waste is generated, and how much is, or can be recycled?
	What pollutants are generated during the process?
Buildings in use/maintenance	Durability of the material in a specific application
	Life expectancy
Downstream – demolition, disposal and recycling	Pollutants caused during demolition
	Pollution as a result of disposal
	The volume of waste being disposed
	The distance material has to be transported either to the tip or recycling point
	What proportion is recyclable or reusable?
	Ease of disassembly

series of checklist of materials to be avoided such as CFCs in air conditioning and insulation or volatile organic compounds (VOCs) in paint. In some cases limited performance dimensions such as embodied energy were included.

These methods struggled to adequately address complex issues such as the contribution they might make in enhancing the performance of the whole

Table 6.2   Current environmental issues

Global issues	Local issues
Global warming	Contaminated land
Acid rain	Solid/liquid waste and landfill
Ozone depletion	External air quality (nitrous oxide, sulphur
Deforestation	dioxide, particulate matter, etc.)
Loss of bio-diversity	Water quality (drinking water and
Resource depletion	watercourses)
	Ecology and bio-diversity, flora and fauna
Health issues	Desertification
	Deforestation
Sick building syndrome	Access to 'green' space
Asbestos/fibrous materials	Noise (from building process and from
Indoor air quality	adjoining owners or users)
Volatile organic compounds	Radon
Drinking water quality	Electro-magnetic radiation

building. An example of this is thermal insulation, which may cause pollution or is energy intensive in manufacture, but reduces energy consumption throughout the life of the building.

The latest versions of BREEAM overcomes some of these problems with a more intensive assessment of materials used in the elements of buildings and the BRE ENVEST2 programme demonstrates what can be done when LCA data underpin a whole building performance tool.

6.11 The future and sustainable building development

To place the discussion above in context, it is important to view it in terms of sustainability issues. The development, use and maintenance of buildings result in a consumption of 6 tonnes of building material per person per year in the UK (BRE 1996) and it is estimated it is 10 tonnes in the USA.

Predictions of the resource efficiency gains that need to be achieved for society as a whole vary from Factor 4 (75 per cent reductions) to Factor 20 (90 per cent reductions). This may seem impossible, but projects have shown some elements such as energy or water efficiency can achieve Factor 4. The Dutch government has instigated research to explore how their construction industry can reach Factor 20 by the year 2050. In 2008 the UK government set the target of cutting CO_2 emissions by 80% by 2050.

The BRE Environment Office has achieved 30 per cent energy reductions and has a 90 per cent recycled content. The Vales have built buildings with an 85 per cent reduction in energy use from that used in buildings built to the Building Regulations prior to the recent revision and their own house in Nottinghamshire is autonomous from the normal external building services.

There is no doubt that we have the technology and expertise when applied together to construct buildings to Factor 4. Whether the political will exists is another story. The other problem is what to do about existing building stock?

References

Ashworth, A. (1999) *Cost Studies of Building*, 3rd edn, Longman.

BEES (Building for Environmental and Economic Sustainability) (1998) *Technical Manual and User Guide*, National Institute of Standards and Technology.

Building Research Establishment (BRE) (1996) *Buildings and Sustainable Development*, Information Sheet A1, Garston.

Building Research Establishment (BRE) (2000–1) *Environmental Profiles of Construction Materials and Components*. Available at http://collaborate.bre. co.uk/envprofiles.

Building Research Establishment (BRE) http://envest2.co.uk.

Curwell, S., Fox, B., Greenberg, M. and March, C. (2001) *Hazardous Building Materials – A Guide to the Selection of Environmentally Responsible Materials*, 2nd edn, E&FN Spon.

Dickie, I. Howard (2000) *Digest 446 Assessing Environmental Impacts of Construction*, Building Research Establishment

Edwards, S., Bartlett, E. and Dickie, I.(2000) *Digest 452 Whole Life Costing and Life-Cycle Assessment for Sustainable Building Design*, Building Research Establishment.

Ferry, D.J. and Brandon, P.S. (1999) *Cost Planning of Buildings*, 7th edn, Blackwell Science.

Flanagan, R., Norman, G., Meadows, J. and Robinson, G. (1989) *Life Cycle Costing: Theory and Practice*, BSP Professional Books

Kelly, J. and Male, S. (1993) *Value Management in Design and Construction*, E&FN Spon.

Kirk, S.J and Dell'isola, A. (1995) *Life Cycle Costing for Design Professionals*, Kingsport Press

RICS (1986) *A Guide to Life Cycle Costing for Construction*, Surveyors Publication.

RICS (1992) *Life Expectancies of Building Components*, RICS Research Paper No. 11.

Stern, N. (2006) *Stern Review on the Economics of Climate Change*, HM Treasury.

Value management or engineering

7.1 Introduction

Value engineering is about taking a second look at key design decisions usually concerned with making savings in the initial capital cost of the building. It does this by looking at the function of an element or component and looking for alternative methods to achieve the same functional performance. With the trend to new types of procurement routes such as construction management, and management contracting coupled with a desire (post-Latham and Egan) to co-operate and work as a team, value engineering has become a valuable tool in improving the final design and construction solution for the building. It is assumed by many that this process is a recent idea, but as the quotation below seems to indicate, perhaps it is not

> When we mean to build,
> we must first survey the plot, then draw the model:
> and when we see the figure of the house
> then must we rate the cost of the erection;
> which if we find outweighs ability,
> what do we then but draw anew the model
> in fewer offices, or at least desist
> to build at all?
> William Shakespeare (1598) *Henry IV, Part 2* Act 1(iii)

Value is not necessarily about money or based upon quantified data, as is illustrated in section 7.3.4, but more about perception of value. What is valuable to one person is not necessarily valuable to another. For instance, the first present a husband buys for his wife may have little monetary value but is sentimentally irreplaceable. A passionate environmentalist will be more concerned about the energy saved using low-energy bulbs than the type of light it emits. Further, there can be several parties involved with conflicting objectives such as hospital trust managers, doctors, nurses and patients. It is the job of the value engineering manager to resolve these difficulties.

7.2 Functional performance

McGeorge and Palmer (2002) identified four levels of function:

*   *Defining the project as a whole.* This is asking very fundamental questions about the purpose of the project. For example, the proposition is to widen a road in the centre of a city to relieve congestion. Superficially the problem is the congestion and widening the road is but one solution. Diverting the traffic onto another route, introducing a one-way system or introducing congestion charges could also resolve the problem. However the fundamental question is: why is there congestion?
*   *Defining the spaces within the project.* The function of the foyer area of a theatre might be to allow the audience to meet and congregate before the performance, stretch their legs during the interval, have a drink and access the toilets. However it might also be used for art exhibitions, a meeting place, providing light refreshments during the day, and workshops.
*   *Defining the function of the elements.* What is the function of a window? Initial thoughts are that it is to act as a ventilator or to admit natural light. However, both of these functions can be achieved by alternative solutions. Light might be provided using a roof light and ventilation provided artificially. It is also there because building occupiers wish to see the world outside and indeed there is evidence to demonstrate that lack of this external link can be mentally unhealthy over a period of time. The window is also part of the exterior wall and its contribution to the overall thermal performance of the outer skin has to be taken into account, and so on. Construction methodology should also be considered. Can the window be fixed in position from the inside of the building thereby eliminating the need for external scaffolding and does this have a knock-on effect in determining how the rest of the outer skin is constructed? What should its dimensions be and how high above floor level should it be positioned. For example, in a hospital ward there are issues of privacy relative to people outside the building.

- *Defining the function of components.* This entails looking at the element's component parts and seeing how they function. Can windows be cleaned safely from inside? Can they be opened readily? Are they childproof? What standard of security is required? What performance is required from the glazing? How is the window fixed to the rest of the element?

7.3 Methodology or sequence of events

Miles' (1972) way, or methodology, of value management has become widely accepted. It comprises seven stages adapted for the purposes of the construction and design processes: orientation, information, speculation, analysis or evaluation, development, selection and conclusion.

7.3.1 Orientation

After the design team has been selected, a meeting is called for all the interested parties. These include the client's representatives, project management and the design team. The prime aim is for all involved in the project to understand the issues and constraints and to set the scene. It also enables those involved to get to know each other and exchange information and ideas. The objective of the meeting is to address four questions:

- What is to be accomplished?
- What the client actually needs? Which is not necessarily what the client thinks it wants.
- What are the key and desirable characteristics of the project?
- A priority list of the key characteristics.

7.3.2 Information

All of the information about the project is gathered together. The accuracy and depth of this process is directly related to the quality of the decisions that can be made and is extremely important. However, this process is also determined by the amount of time available to collect it. Facts are more important than assumptions. Initially the kind of information required is to provide answers to the following:

- *Client's needs.* These are the fundamental requirements which the project must satisfy. This may not be just the physical needs of the building in satisfying specific functions, but may also include any statement the building must make for the client. For example, it is the head office of a major company or organisation, a national symbol or the greatest shopping experience in the area. The National Westminster Tower, Lloyds of London, Coventry Cathedral, The Welsh Assembly building and the Trafford Centre are examples of this. In projects not value managed, it is not unusual for decisions to be made without detailed consultation with those persons who are actually going to work or use the space. At some stage in the process, this level of consultation needs to take place. In other words, the client is not just the person or persons who commissioned the building.

- *Client's wants.* These are things the client might want but are not essential to the function or the statement of the building.
- *Project constraints.* These are issues outside the client's or designer's control, such as planning restrictions, shape of the site and rights of way.
- *Budgetary limits.* The amount of money the client has to spend on the land acquisition and the building. This will also involve decisions about life-cycle assessment issues.
- *Project duration.* This includes the design and construction phases. A retailer may stipulate they must have the building by September, so they can be in business for the Christmas rush.

What should also be happening at this stage is the commencement of team building as this is the first time the main participants are all brought together. If this is thought out properly, the advantages to the rest of the process and project will be greatly enhanced. As the process continues more detailed analysis can take place at the differing levels of function (section 7.2).

7.3.3 Speculation or creativity

Brainstorming is a popular way of creating new ideas and resolving problems, although not the only way. The group is asked to consider the problem and ideas are recorded and considered even if, on the surface, they appear ill-conceived or inappropriate. What often happens is that an 'ill-conceived' idea develops a different train of thought that leads to the resolution of the problem. The group should be encouraged to generate a large number of ideas and not to ridicule any submitted.

An example of brainstorming in practice was in the manufacture of the Jespersen industrialised building precast floor components, similar in design to today's Bison floor units. The components were made on a conveyor line, and the whole ethos of the production process was that any insert or former used to produce a non-standard component, would be positioned just prior to casting and be removed directly after casting had been completed. The mix design and severity of vibration permitted this to happen. The horizontal holes were made using withdrawable steel tubes. This worked well except for the forms used in the mould ends when it was necessary for one or more of the tubes used to form the horizontal holes through the concrete unit to be withheld to stiffen the component and act as an edge beam. These formers kept vibrating out.

Staff were assembled to discuss the problem. One of the foremen from another department suggested that cardboard forms should be placed inside the mould stopping the concrete from coming out. The idea of cardboard was impractical, as it would have to remain in the mould until the concrete had cured sufficiently for it to be extracted. It would have been difficult to remove from the cured concrete, as it would have adhered to it. However, the idea of blocking from the inside, resolved the problem. The solution was to insert cylindrical timber bungs into the mould ends, with a small lug to stop the bung from sliding through the hole. These remained during the casting and curing process, and the problem was resolved.

Evidence seems to indicate, as in the example, that it is the inexperienced member of the group, rather than the expert who is more likely to come up with the most original suggestion, probably as that person is not restricted by tradition and convention.

7.3.4 Analysis and evaluation

The team looks at the various suggestions, eliminating the ones that have no place because they may not fully satisfy all the functional requirements, are unsafe, impractical, or far too expensive. This is a relatively simple stage and acts as a filter for the next. There are a variety of ways of doing this but a simple and often used technique is to rate each design criterion for each appropriate solution. The rating scale chosen for this example is: Poor = 1, Fair = 2, Good = 3, Very good = 4, Excellent = 5.

Applying these ratings to various types of floor finishings might produce answers as shown in Table 7.1. The row entitled 'weighting' is the designer's/client's perception of importance of the various criteria. For example, if the view is that environmental impact is the most important then 5 might be used. Initial costs may be considered more important than aesthetics, or vice versa. Using the same rating scale, the figures in the rows showing the various finishes are the designer's assessment of the performance against

Table 7.1 Sample evaluation of flooring finishes

Type of finishing	Initial cost (5 = cheapest)	Aesthetics	Cleaning	Replacement frequency	Environmental impact (1 = greatest)
Weighting	3	5	3	2	3
Flexible PVC tiles	3	3	4	4	3
Terrazzo	2	4	5	5	4
Wood blocks	1	5	3	5	4
Nylon carpet tiles	4	3	3	2	2

Table 7.2 Sample solutions for various types of flooring finishes

Type of finishing	Initial costs	Aesthetics	Cleaning	Replacement frequency	Environmental impact	Total
Weighting	3	5	3	2	3	
Flexible PVC tiles	3×3	3×5	4×3	4×2	3×3	53
Terrazzo	2×3	4×5	5×3	5×2	4×3	63
Wood blocks	1×3	5×5	3×3	5×2	4×3	59
Nylon carpet tiles	4×3	3×5	3×3	2×2	2×3	46

each criterion. The figures in the table reflect subjective opinion and should not be taken as definitive.

By then multiplying the weighting of importance by the performance rating given and then adding these up, a guide is given as to which is the most appropriate solution for the client's needs, as shown in Table 7.2.

It can be seen by in this case that terrazzo tiles are the most favourable with the highest score, followed by wood blocks, but it must be remembered that in this example the weighting is based on the writer's subjective judgement only. Further both the weighting and rating scale of 1 to 5 is crude, but this can be improved by having a broader scale.

7.3.5 Development

This stage takes the solutions still remaining after analysis and evaluation and develops them in some depth, which also includes buildability issues. Outline designs and costs are produced for each design. Life-cycle assessments will also be produced at this stage if it is appropriate to the developer's long-term plans. Those that do not meet the prime design criteria are eliminated. Some further thoughts on methodology are given in sections 7.5 and 7.6.

7.3.6 Selection

The various solutions left from the previous stage are now presented to the management team and client for a final decision to be made. Drawings, methodology statements, programme implications and costings will be used to support these solutions. A commentary on the benefits or otherwise can also be included.

7.3.7 Conclusion and feedback

The client makes the final selection from this information. What is also important is that on completion of the building works, a feedback loop is completed to enable lessons to be learnt for the future.

7.4 When valuation engineering should be applied

The later in the design and construction processes value engineering is applied, the less likely it is that money will be saved as the potential for making change is reduced. The costs of implementing change also increases as demonstrated in the Figure 7.1. A simple and obvious example to clarify the point would be that it is easier to move a column on the drawing than it is to move it once constructed. More significantly, by applying value engineering to the footprint of the building or allocation of space within the footprint, significant economies can be made in terms of the capital cost of the building, staff resources needed in maintaining the building and life-cycle costing issues.

7.5 Functional analysis method

In functional analysis the function of each component or process is examined by asking the question 'What does it do?' followed by a second question, 'How else can this be achieved?' For example: 'How is paint applied to the wall?' Answer: 'With a brush'. Or, 'How else can this be achieved?' Answer: 'Using a roller or spray'. These assume the wall is built in situ. If the wall is a factory-made component, then alternatively the wall could be dipped.

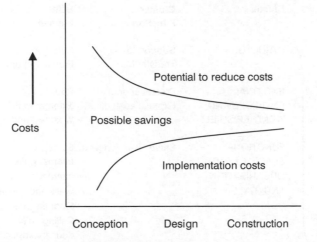

Figure 7.1   The timing of valuation engineering

However, this does not go far enough because the function of the wall surface has first to be questioned. What does it do? Depending on the application it could need to be:

- resist impact
- hygienic
- easy to clean
- decorative and colourful
- water resistant
- resistant to UV light
- suited for certain acoustics
- flame and fire retardant.

Deciding what the function of the surface is determines the types of paints appropriate for the function which then determines the method of application. The choice of surface finish might then determine the surface and material required to receive this paint and vice versa. This might be an expensive combination so it would be necessary to go round the loop again. In the end though the finish must satisfy the function desired from it.

7.6 Function cluster groups

There have been found to be benefits in dealing with clusters of functions. The BICS (Building Cost Information Service) standard elements have

Cluster	Cluster Function	BCIS Element
STRUCTURE	Support load	Substructure
	Transfer load	Frame and upper floors
EXTERNAL ENVELOPE AND WEATHER SHIELD	Protect space	Roof
	Express aesthetics	External wall
		Windows and external doors
FUNCTIONAL SPACE, CLIENT ORGANISATION AND STYLE	Serve client function	Stairs
		Internal walls and partitions
		Internal doors
		Walls and finishes
		Floor finishes
		Ceiling finishes
		Sanitary appliances and plumbing
		Communications and data
INTERNAL ENVIRONMENT	Maintain comfort	Space heating
		Ventilation and AC
		Electrical
INTERNAL TRANSPORTATION	Minimise walking	Lift and escalator
	Raise/lower people and loads	
EXTERNAL INFRASTRUCTURE	Protect property	Site works and drainage
	Circulation/parking	external services
	Remove waste	

Project mission and function of building { (bracket encompassing all cluster rows)

Figure 7.2 Function cluster groups

been used in Figure 7.2. (reproduced from Kelly *et al.* (2002) *Best Value in Construction* with permission of Blackwell Publishing).

The advantage of this approach is that it focuses the mind to specific areas, readily understood especially by the client. The client may be particularly concerned with overall issues, such as the appearance of the building, the quality of the internal finishings or environmental impact of the building. In each of these cases the function cluster assists in this process. It can be from this approach that the value engineering can be directed in to specific issues more readily than taking individual components.

References

Kelly, J. and Male, S. (1993) *Value Management in Design and Construction*, E&FN Spon Press.

Kelly, J., Morledge, R. and Wilkinson, S. (2002) *Best Value in Construction*, Blackwell Science

McGeorge, D. and Palmer, Λ. (2002) *Construction Management New Directions*, 2nd edn, Blackwell.

Miles, L. D. (1972) *Techniques for Value Analysis and Engineering*, McGraw-Hill.

Procurement methods and types of contract

There was a time when procurement was very much prescribed. The client would appoint an architect who would select the rest of the design team. This would include the quantity surveyor and structural engineer. If the client was experienced, they may have employed the other members of the design team directly. Bills of quantities were produced as explained in Chapter 9. Contractors would tender for the work, and the lowest price, especially for public-sector work, determining who would be awarded the contract. Contractors worked to the letter of the contract using every clause available to obtain extra payment. The result was often confrontation between the design team and the contractor, the client picking up the bill at the end.

This has changed over the last few decades with a multitude of different ways being developed for clients to procure work. Many of these changes have occurred because of clients' dissatisfaction at the way the industry was operating. A significant turning point was the publication of the report *Faster Building for Industry* (1983) that investigated different types of procurement then in use and analysed the effectiveness of each on a large number of contracts. The conclusions were interesting in that the traditional methods of procurement were not as successful as others in use, especially those that were more client orientated. This report anticipated many of the later findings of Latham (1994) and Egan (1998).

8.1 Traditional methods of procurement

8.1.1 Competitive tendering

This method is outlined in the introduction and is sometimes referred to as single-stage selective tendering. A significant amount of design has to be completed before the quantity surveyor can produce the bills of quantities (B of Q) and the contract is put out to tender. Some work is offered on open tenders permitting any contractor to apply, but this can mean far too many contractors apply than is manageable. Pricing a contract is costly and this approach causes a considerable amount of wasted effort by the numerous tenderers. Much work is offered as selective tendering, when only contractors who have the prerequisites to carry out this work are invited to tender. Many clients operate a selective tender list to which contractors can apply for inclusion. This will involve interviews and presentations allowing the contractor to demonstrate its ability to carry out work for the client.

On being awarded the contract, the main contractor puts the sub-contracts and materials requests out for tender, unless the architect has already nominated them in advance. This occurs occasionally when, for example, an order is placed in advance because the lead time for the manufacturer is such that delaying the placing of the order could cause delays during the construction process. As can be seen in Figure 8.1, the duration from appointment of the architect to commencement on site is lengthy.

The dotted arrows show the contractual links and the solid lines the control, i.e. management links, between the various parties. The contract used in this case is almost invariably the JCT05 Standard Building Contract with Quantities. There can be other links between the client and specialist sub-contractors if they carry out a design function. The B of Q is used as the basis for the financial administration of the project (Chapters 13 and 15) and any errors in the pricing is borne by the contractor. Recompense

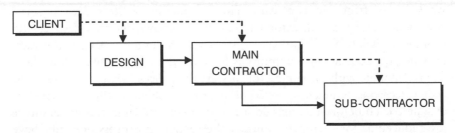

Figure 8.1   Traditional method of procurement

for any delays or mistakes made by the design team, are claimed for by the contractor.

It is important to note that the main contractor has little or no effect on the costs of the site selection, design, life-cycle costs or impact on the environment as a whole, so all their experience and expertise is lost. The competitive tendering process requires the production of a combination of the following information for the contractor to price the work:

- specifications and drawings
- performance specifications
- bills of quantities
- bills of approximate quantities
- schedules of rates.

8.1.2 Negotiated tender

In this method, the contractor is selected early in the process which enables their expertise to be used in the design process. Often several contractors are approached and asked to submit costs for the management of the project, key items of work and profit margins. This approach has been more finely tuned for prime costing contracts (section 8.8). One of the advantages of this method is that the work on site can commence earlier than the competitive tender process permits. This is sometimes referred to as a two-stage process. Bringing the contractor in early enables it to make a technical contribution to the design, improve buildability, modify the design to suit specialist plant and equipment, order materials early to reduce delays, and above all, bring in sub-contractors to contribute in a similar way. Finally they can bring their management skills into the process. The contract between the client and various parties can be adapted to suit the particular needs of all.

It is ideal when the client has a programme of new or refurbishment work to be carried out over a period of time as it enables the same parties involved – the designers, client and contractors – to build up an understanding and work more effectively as a team. This is sometimes referred to as serial contracting and the contractor tenders in the knowledge that if successful more work will be forthcoming. A further variation on this is term contracts in which the contracting continuity is for a fixed term after which the client can renegotiate with the contractor or start the process again. In these cases, cost forecasting is much more accurate for the client as the cost data are available even before the next contract commences. Time and expenditure are reduced as for each new contract the tendering process does not have to take place. Opponents of this method argue the client doesn't obtain the lowest price, but then does the lowest price always provide the best solution for the client? Also there is less likelihood of claims being submitted at the conclusion of the contract. The contractual relationships remain the same as outlined in Figure 8.1.

8.1.3 Competitive tender based on approximate bills of quantities

Using the same contractual relationships, an alternative approach is to use approximate quantities. This method is used when the project has not fully been designed and a full B of Q cannot be produced in time for the tendering process to take place. As the contract proceeds, work is measured on a monthly basis and the monies paid are agreed based upon the approximate B of Q as far as is possible. This permits the construction work to commence earlier, but financial control can be less accurate than with a full B of Q.

8.2 Design and build

This is usually a competitive process, but can be by negotiation. The client invites contractors to tender not just for the construction work but also for the design giving an inclusive price. This can be a fixed price, which clients find attractive, as one of the major variables in their budget is eliminated and there can be no claims for late design information. However, there can be some difficulty in selecting a contractor because comparisons between the different designs, costs and time required for the execution of the design and construction may vary. It is important the client selects a contractor who specialises in the type of building required. After selection they only have to deal with one organisation compared with the traditional method.

The client should write a comprehensive brief to suit their requirements otherwise there is the danger that the contractor will produce a building more inclined to their production needs rather than the needs of the client. This places the inexperienced client at a disadvantage as they may find difficulty in doing this.

In the early days of design and build many contracts were for factory sheds, however this has now developed into such common practice that major contracts are procured this way. The advantage of this method is the contractor has specialist knowledge that may outweigh that of the architect normally employed by the client, resulting in more cost-effective solutions and fewer variations. Compared to traditional methods it provides a more rapid completion of the project as the design and construction processes can

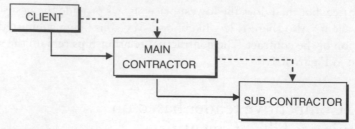

Figure 8.2   Design and build

overlap. The contractor can use its own design team or designers they regularly work with, and can tailor their production expertise and produce buildable solutions. In some cases the client will have produced outline drawings and performance specifications upon which the contractor will develop the initial design and then construct it. They may employ an agent to supervise the contractor, especially if they are inexperienced. The advantage of doing this is that the client maintains a design overview and can more significantly influence the final design outcome at a stage when it is easier to change.

Another form of design and build contract is turnkey when the builder not only designs and constructs the building, but also provides all that is necessary for the client to commence their activity. All the client has to do is 'turn the key' in the front door. The contract normally used for this type of procurement is the JCT05 Design and Build Contract. In Figure 8.2 the dotted arrows show the contractual links and the solid lines the control between the various parties.

The client loses some control over the project as the risk is being largely borne by the contractor. This means also that any alterations the client may require can be more expensive than using the traditional method, although increasingly contractors try to accommodate the client's needs whenever possible.

8.3 Construction management

There can be some confusion between the terms construction management and management contracting (section 8.4). So for clarity both sections commence with a definition for the terms used in this text. In this case the client employs the design team and then a construction manager is employed to programme and co-ordinate the design and construction processes for which they are paid a fee. The role also involves improving the buildability of the design. Figure 8.3 demonstrates the contractual relationships, the dotted lines contractual, and the solid lines control responsibility.

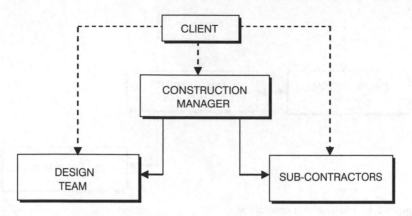

Figure 8.3   Construction management

The construction management team does not carry out any direct work themselves, but divides the work into packages and sublets them to specialist sub-contractors. The contract is between the client and the sub-contractor even though the construction manager will have usually negotiated the work package on behalf of the client. This means the client has to be very knowledgeable; if not, this is not an appropriate method of procurement. The design team will be outsourced totally, either selected in part or whole by the client, or totally by the construction management organisation.

The term design team is a loose description as within the process of development it may well include many other professionals, such as lawyers and financiers (Chapter 1). This is because the process of management can commence as early as the feasibility of the project. A good example of this is when a client is competing for a PFI contract and employs a management contractor to control the process.

The construction phase can commence prior to the design being completed, and because of the nature of the contractual relationships, the traditional adversarial conflict between the various parties is removed. It is also possible to change the design as the contract progresses providing the particular work package has not been let and it does not have an effect on packages already awarded.

Unlike design and build contracts, where the client knows reasonably accurately what the final cost of the building is likely to be, in this case until all the packages are let there remains uncertainty and projected costs rely on the elemental cost plan the private quantity surveyor (PQS) has produced. This means the construction manager has to be very cost-aware to be effective.

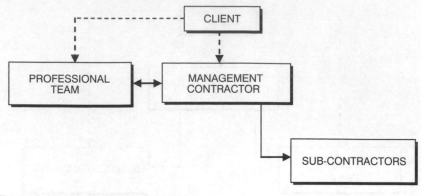

Figure 8.4 Management contracting

8.4 Management contracting

Here, a contractor is employed by the client to manage the construction process and is paid a fee. The two main differences compared with construction management are first, the contract for the sub-contractors is with the management contractor rather than with the client and second, whilst having a significant input into the design process, the management contractor does not have the responsibility to manage this part of the process.

8.5 Partnering

There are two types of partnering agreements: that of strategic or multi-project partnering, and project partnering which is for one project. It may well be that a successful project partnership leads to a strategic agreement. The greater benefits accrue on multi-project partnering as this allows the partnering arrangements to be extended to the supply chain (*Operations Management for Construction*, Chapter 7). When this occurs it is a logical extension of continuity type contracts such as serial contracts (section 8.1). However, it should be noted that for EC and UK public projects, partnering agreements can only take place after the contract has been awarded, which makes strategic partnering difficult to use.

The essential aim of partnering arrangements is the achievement of trust and co-operation between the partners. It is a commitment of all the partners to make the project successful and it is this attitude, rather than the contractual links that drives the relationship. It is argued that this commitment cannot take place unless the decision to enter a partnership is taken at the highest level in all the participating organisations. The partnering process is about team building, communication and team spirit so all involved work together as one, towards the common goal as the prime objective, eliminating personal organisation interests from their thinking. For those interested in implementing partnering, see Chapter 8 in McGeorge and Palmer (2002). Partnering should commence at the development and design stage and it is important that, if possible, the composition of the team should remain stable throughout the duration of the project.

In achieving the proper relationships, it is necessary to agree three key objectives as shown in Figure 8.5: mutually agreed objectives, such as improved performance and reduction in costs; actively looking for continuous improvement; and having a common approach to resolving problems.

It is important to draw up an agreement between the partners. This should cover the following:

- a statement of working in good faith;
- a method by which open-book costs of all the partners can be demonstrated and seen by all parties;
- a clear statement of the roles and responsibilities of those involved;
- how the lines of communication between parties will work;
- a procedure for resolving disputes. Hopefully this will not have to be enacted, but it will need to be clearly defined.

The client in the partnership agreement is not necessarily from one of the traditional property development companies and one of the earlier examples

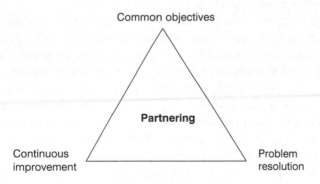

Figure 8.5 Partnering (adapted from The Aqua Group, *Tenders and Contracts for Building*)

of partnership in construction demonstrates this, as described in the brief case study below.

The Barnsley Partnership

The Barnsley Metropolitan Borough Council (BMBC) made the decision the form a partnership. They were seeking input from private enterprise in a rolling programme of development using sites owned by the council. These were not to be selected because they were the most lucrative to the partner to develop. They then ran a competition to find a contracting partner and Costain Construction was the successful candidate. The partnership was formed in 1990.

All profits and losses were shared equally between the partners, which included a proportion of the contracting profits. The company offered a social dimension to the partnership by way of integrated training programmes for local people and also participated in local community activities. One example of this was that the company's human resources department worked up and sustained a youth centre in one of the deprived parts of the town. Another was working with Kwiksave to provide staff training. A demonstration of the success of this aspect of the partnership was the award of the Lord Mayor of London's trophy for participation in the community.

Within the terms of the agreement, Costain was allowed to negotiate the price on three out of five projects, allowing the BMBC to tender on the other two for the purposes of maintaining Costain's competitive edge. The negotiation was not a pushover for Costain in that the negotiator from the council was not part of the partnership and was in close touch with the current construction costs protecting the council's interest.

The expected output for the council was to use the expertise of the private sector as well as the company's client base to improve the local economy and environment, dispose of some difficult sites, and earn some extra money. On the other hand Costain expected to gain from the development and construction profits from the non-competitive projects and get the opportunity to win similar types of partnerships in other parts of the UK. It was also felt that at that time doing business with a social dimension was a valued cachet, especially by some of the trade unions which had been influential in the concept and award phases. It was also perceived the partnership would benefit from having an understanding planning department in obtaining planning permission.

On the surface the idea that a strongly socialist council could have a partnership with a private company perceived as being more 'right wing' was an interesting concept. In practice, clashes of culture were never allowed to interfere with the management of the partnership. The agreement was a tightly worded document and the management board was charged to ensure this aspect of working together would not be a problem.

A Costain employee was appointed as the chief executive of the partnership, and reported to the management board that consisted of the council leader, the chief executive of the council and the director responsible for the valuation and planning department.

The partnership had its successes, but also disappointments. However others saw the business as a model of success such that an adjoining borough approached the chief executive with a view to setting up a partnership there. This was declined, for such circumstances had been envisaged in the articles of agreement, which specifically disallowed setting up a partnership with an adjoining borough.

Benefits accruing from partnering in general include:

- by definition it eliminates the adversarial approach between the partners and this minimises the likelihood of incurring legal costs;
- working relationships develop over the long period of association, although if not watched this can lead to complacency;
- in strategic partnering, tendering costs are reduced since each new contract awarded is not done in competition with others;
- opportunities for innovation and value engineering;
- efficiency occurs as a result of familiarity with each other's systems, which may also be modified to assist in supply chain management;
- the anticipation and identification of risk is improved.

On the negative side if not properly controlled and monitored:

- personal interests can take precedence over the common good;
- new personnel do not have the same enthusiasm as those who were in at the beginning and hence do not buy into the philosophy;
- in the event of dissolution of the partnership, parties revert to type and look out for their own interests at the expense of others;
- lack of competition results can in higher costs.

8.6 Alliancing

This is really a development from partnering and is more encompassing. At the time of writing there is only a limited amount of experience of it in practice in the construction industry, but it is felt the reader should be aware it is a form of contract that may become more commonly used.

8.7 Private–Public Partnerships and Private Finance Initiatives

The basic idea of the Private Finance Initiatives (PFI) is that private companies build the project, such as a hospital, school or road using their own money normally provided by the banks (Chapter 2). The government rents it back from them over a given period after which the ownership of the building either returns to the government or a further contract is negotiated. The PFI became very complex and in 1997 the Bates Review made 29 recommendations to streamline the process. This became known as the Private–Public Partnerships (PPP).

Many governments around the world are increasingly adopting this form of partnering as a means of reducing the strain on the public purse, the desire to embrace best practice, using key performance indicators, achieving best value and the need to produce transparent audit trails. In the UK the amount of PFI work expressed as a percentage of the total capital spending on construction increased from less than 1 per cent in 1994 to 16 per cent in 2001 (*The Economist*, 15 September 15 2001) and has been supported by the main political parties.

Some contractors have been put off the PFI because of the high costs of bidding and the inherent financial risk. This is because if, after the initial interest shown, the contractor is short-listed, a significant team of experts has to be put together, all of whom generally work for no remuneration but rely on their costs being recouped if they are successful with the bid. The bidding process can be very costly as it will usually include providing a design solution to meet the client's brief, carrying out value engineering on the solution and deciding on production methods, and then agreeing sources of finance. All of this takes time and money and on major contracts could run into millions of pounds.

8.8 Prime costing

This is an extension to design and build contracts (section 8.2). Other than the work that has to be done initially prior to selection for which no fee is paid, this is risk-free type of contracting and is therefore very attractive to the contractor. Usually about six companies are approached to produce a pre-qualification document. After consideration the client then reduces the number to about four companies. In competition the construction company puts together technical proposals, a design solution and an indication of the costs, which means assumptions have to be made since the final design has not been agreed. A preliminary cost plan is also produced along with an estimate of the preliminary costs.

After this stage, two contractors are selected for the final phase of the process. These two then make presentations to the client, and after client evaluation, the preferred bidder is chosen. From this stage the contractor is reimbursed for all work. What is important is there is transparency on all costs, including the supply chain. So a sub-contractor doing the cladding would advise a price for fixing the cladding, but would itemise this into cladding as an item, the insulation added, the labour and any plant required. The contractor works closely with the client and the end users, if appropriate, to develop the design. Examples of end users in a hospital development would be the doctors, nurses, administrators and patients. It means fully understanding the work process in the proposed development, the number of people employed in each section of the process and any requirements they need. The design is then produced. This also involves working closely with the suppliers and sub-contractors on their design if they have been commissioned to do this.

8.9 ProCure21

Similar to prime costing this is a form of contract trust hospitals have been using and owes its origins to the report *Rethinking Construction*. It is important to be aware of this form of procurement because of the major investment being made into new hospitals. Prior to this, the National Health Service (NHS) almost invariably procured on the traditional adversarial approach and looked solely for the lowest cost. The aim of the NHS was to promote better capital procurement, by taking note of the best practice in the private sector, and by benchmarking using key performance indicators. After a pre-selection process similar, but more intense, to that described in prime costing (section 8.8), five construction companies were originally selected, known as 'supply chain members', but this number has now been expanded. Each company agrees its overheads and profit margins even though they may differ from company to company. When a trust has new work it can approach any of the approved contractors which suits their needs the most, which saves going through the time-consuming and expensive process of tendering.

The main benefits of the scheme are that the supply chain members have already been selected, so the procement process can save up to 12 months as it is not necessary to go through the *Official Journal of the European Union* (*OJEU*) process for individual schemes. The fact supply chain members are involved in the early design decisions means the design process is faster allowing the contractor to get onto site earlier. It also means that as a result of carrying out several contracts the contractors become more aware of their client's requirements so design decisions are improved. This means the client obtains an improved building meeting their needs more, and there are continuous innovations and improvements, resulting in reduced costs for the client and increased profits for the contractor.

8.10 Apportioning risk

HM Treasury defines risk as 'the uncertainty of outcome, whether a positive opportunity or negative impact'. In development, risk is the difference between the anticipated cost of the building, including all development, design and construction costs, and the final actual costs. Since a significant factor in the procurement of contracts is concerned with distribution of financial risk between the various parties involved, a summary of this has been produced in Table 8.1. It should be noted that the more the client

Table 8.1   Financial risk

Type of procurement	Risk
Traditional	The contractor takes a risk initially when tendering in competition. Once the contract is awarded, the client carries the risk.
Design and build	Contractor takes the risk and responsibility. If this is a fixed sum rather than having some form of inflation-proof contract, the risk is greater.
Construction management	Depends upon the form of contract, but generally the contractor takes the greater risk, but since packages are let closer to execution of the work, less risk is built into the package price by the sub-contractor.
Management contracting	Same as construction management.
Partnering	Partnering is not about allocation of risks but, as Latham asked for, 'shared financial motivation'.
Alliancing	Similar to partnering.
Private Finance Initiatives	The unsuccessful bidders can take considerable financial risk, after which some of the risk is transfered to the client, the amount depending upon the wording of the contract.
Prime costing	Other than during the pre-qualification period, the client takes the risk.
ProCure21	As with prime costing client takes risk and responsibility.

apportions risk, generally speaking, the less control the client has over the project. For simplicity the risk apportioning shown is between the client and the contractor. However, the client will have almost certainly laid off some of the financial risk with the financers of the project. Much depends upon the way the contract has been drafted so the table is only a rough guide.

References

Aqua Group (1999) *Tenders and Contracting for Building*, 3rd edn, Blackwell Science.

Egan, J. (1998) *Rethinking Construction*, HMSO.

Ferry, D.J. and Brandon, P.S. (1999) *Cost Planning of Buildings*, 7th edn, Blackwell Science.

JCT05 Design and Build Contract (2005), RIBA Publications.

JCT05 Standard Form of Building Contract with Quantities (2005), RIBA Publications.

Kelly, J., Morledge, R. and Wilkinson, S. (2002) *Best Value in Construction*, Blackwell Science.

Latham, M. (1994) *Constructing the Team*, HMSO.

McGeorge, D. and Palmer, A. (2002) *Construction Management New Directions*, 2nd edn, Blackwell.

National Economic Development Office (1983) *Faster Building for Industry*, HMSO.

NHS ProCure21 www.nhs-procure21.gov.uk (accessed 12 January 2009).

Smith, R.C. (1986) *Estimating and Tendering for Building Work*, Longman.

Standard Method of Measurement and bills of quantities

9.1 Introduction

The purpose of this chapter is to introduce the bills of quantities and the Standard Method of Measurement, so that their composition, application and usage are understood. It is not intended to teach the reader how to produce bills of quantities in practice.

In the past much of the contracting work was obtained via the traditional route as described in section 8.2 using bills of quantities. Whilst in recent years the amount of work using this approach has declined, much of other forms of contract between the client and the contractor are based upon the principles laid down in these two documents. The employer's quantity surveyor produces the bills of quantities. It is an itemised account of all the work that has to be carried out on the contract. The method and sequence of measurement in the bills is based upon the standard method of measurement.

9.2 The Standard Method of Measurement

The great advantage of the Standard Method Measurement, now in its seventh edition (SMM7), is that it determines a uniform method accepted by all parties. Prior to this document estimators were often left in doubt as to what items in the bills of quantities actually meant. This could lead to misunderstanding, inaccurate tendering and a basis for dispute when payments were made for work completed.

In 1912 a joint committee, made up of members of the Royal Institution of Chartered Surveyors and the Quantity Surveyors' Association, was charged with producing a standard set of rules for measuring building works. In 1918 four contractors nominated by the Institute of Builders and the National Federation of Building Trades Employers were added to the committee. From time to time input was sought from representatives of certain trades. The first edition was published in 1922 and was based on the practice of leading London quantity surveyors of the time, modified to suit practices throughout the UK. The most recent edition was published in 1988.

9.3 Composition of SMM7

The Standard Method of Measurement is divided into different categories as shown in Table 9.1. Each is a clearly defined area of work. All of the categories, with the exception of A, have a physical presence in the final building. The preliminaries are discussed in section 9.4. It is important to note that the sequence shown is the sequence in which measured work is

Table 9.1   Detailed contents of SMM7

Categories	Categories
A. Preliminaries/general conditions	P. Building fabric sundries
C. Demolition/alteration/renovation	Q. Paving/planting/fencing/site furniture
D. Groundwork	R. Disposal systems
E. In situ concrete/large precast concrete	S. Piped supply systems
F. Masonry	T. Mechanical heating/cooling/refrigeration systems
G. Structural/carcassing metal/timber	U. Ventilation/air conditioning systems
H. Cladding/covering	V. Electrical supply/power/lighting systems
J. Waterproofing	W. Communications/security/control systems
K. Linings/sheathing/dry partitioning	X. Transport systems
L. Windows/doors/stairs	Y. Mechanical and electrical services measurement
M. Surface finishes	Additional rules – work to existing buildings
N. Furniture/equipment	

presented in the bills of quantities. This uniformity helps the user of the bills assess information.

Each of these sections is subsequently broken down into further subsections as shown in Table 9.2. There is a clear distinction in the types of material being used to aid clarity. It is also useful to have it broken down in this way for the contractor, because if all like materials and processes are together it is easier to abstract information to send to suppliers and sub-contractors for quotations. It can also be used to provide material quantities for the planner to produce the pre-tender programme and later on, for ordering materials.

The SMM7 then takes each of these subsections, or in some cases as shown in Table 9.3, combines them (F10 and F11 from Table 9.2), and

Table 9.2   Contents of section F – masonry

	Sub-section
F10	Brick/block walling
F11	Glass block walling
F20	Natural stone rubble walling
F21	Natural stone/ashlar walling/dressing
F22	Cast stone walling/dressing
F30	Accessories/sundry items for brick/block/stone walling
F31	Precast sills/lintels/copings/features

Table 9.3   Brick/block walling and glass block walling

Classification table F10 Brick/block walling F11 Glass block walling	A	B	C
1. Walls 2. Isolated piers 3. Isolated casings 4. Chimney stacks	1. Thickness stated 2. Facework one side, thickness stated 3. Facework both sides, thickness stated	1. Vertical 2. Battering 3. Tapering one side 4. Tapering both sides	m²
5. Projections	1. Width and depth of projection stated	1. Vertical 2. Raking 3. Horizontal	m
6. Arches	1. Height on face, thickness and width of exposed soffit and shape of arch stated		m

Table 9.4 SMM7 further rules for section F – masonry

Measurement rules	Brickwork and blockwork unless otherwise stated are measured on the centre line of the material
Definition rules	Thickness stated is nominal thickness unless defined otherwise below
Coverage rules	Brickwork and blockwork are deemed to include (e.g. extra materials for curved work)
Supplementary information	Kind, quality and size of bricks or blocks

produces further sub-classifications. At the same time it goes into further detail including whether measured as linear, metre super or metre cube, all of this with a view to clearly specify the finished work. Table 9.3 demonstrates only 6 of the 26 classifications shown in the SMM7. Note the descriptions grouped in columns A and B relate to each individual description grouped in the classification table column. So, a description might be vertical walls, facework one side, 225 mm thick.

Not shown in Table 9.4, but included in the SMM7, are further instructions such as measurement rules, definition rules, coverage rules and supplementary information for each of the categories. An example of each of these is shown in Table 9.4.

9.4 Preliminaries

This is the first section (A) of the bills of quantities. Preliminaries are concerned with the administration of the project and the provision of plant and site-based services such as accommodation, water, electricity, etc. The content will vary considerably from contract to contract, but always needs careful attention. Generally the value of the preliminaries can be from 7 to 15 per cent of the total cost of the project. Errors here can have major implications on the cost performance of the contract as a whole. For example, the salaries of staff employed on the contract are included here and if the contract takes longer to complete than programmed through no fault, there will be no more money left to cover these wages. Equally, if an extension to the contract is agreed, for example to accommodate alterations made by the client, then these figures act as the basis for the calculation of recompense to the contractor and hence the need for accuracy (see Chapter 14).

As the value of the preliminaries is so significant, if contractors were paid for them proportionately over the period of the contract, their cash flow would be severely distorted because of the initial outlay that has to be made to set up the contract for such items as site accommodation, installation of a tower crane and hoardings around the site. To overcome this, preliminaries are paid for in two ways:

- *Fixed charge*. This is for work considered to be independent of the duration of the contract.
- *Time-related charge*. This is work considered to be dependent on duration.

In essence there are five main sections in the preliminaries:

- General project details (A10, 11, 12, 13)
- Contractual matters (A20)
- Employer's requirements (A30–A37)
- Contractor's general cost items (A40–A44)
- Information to be carried out by groups other than the contractor (A50–A55).

The complete list of items is shown in Table 9.5.

9.5 Preliminaries classification expounded

The examples given for the contents of the various classifications following are only meant as an indication of what might be included. In reality, not all may be used and equally they may be expanded upon. Note the SMM7 always refers to the client of the project as the employer.

9.5.1 General project details

A10 – Project particulars

- Name, nature and location of the project. This gives the contractor an immediate overview of the project and sets the work in context. The location is important to know in terms of how far it is from the office which affects communications, the difficulty in recruiting sufficient local labour which could mean transporting labour in, and if it is familiar territory.
- The names and addresses of the employer (client) and consultants. This is important to know as the contractor may have worked with some or all of these parties before. The quality of this experience could affect the decision on whether to tender depending on previous experiences.

Table 9.5   Preliminaries classification list

	Classification		Classification
A10	Project particulars	A37	Employer's requirements: Operation/maintenance of the finished building
A11	Drawings	A40	Contractor's general cost items: Management and staff
A12	The site/existing buildings	A41	Contractor's general cost items: Site accommodation
A13	Description of the work	A42	Contractor's general cost items: Services and facilities
A20	The contract/sub-contract	A43	Contractor's general cost items: Mechanical plant
A30	Employer's requirements: Tendering/sub-letting/supply	A44	Contractor's general cost items: Temporary works
A31	Employer's requirements: Provision, content and use of documents	A50	Work/materials by the employer
A32	Employer's requirements: Management of the works	A51	Nominated sub-contractors
A33	Employer's requirements: Quality standards/control	A52	Nominated suppliers
A34	Employer's requirements: Security/safety/protection	A53	Work by statutory authorities
A35	Employer's requirements: Specific limitations on method/ sequence/timing	A54	Provisional work
A36	Employer's requirements: Facilities/temporary works/ services	A55	Dayworks

If there had been an excellent working relationship and the previous contract(s) have run smoothly, then this could affect the tender price and produce a more competitive bid.

A11 – Drawings

- This is a list of the drawings from which the bills of quantities was produced. This gives the contractor an indication of the stage the design has reached and therefore the reliability of the data on which the tender is being based. The less detail available then the greater the risk of the estimate being inaccurate. This increases the likelihood of claims for extra payments or extension to the duration of the contract, which can sour relationships with the employer as they see the cost of the project rising.

A12 – The site/existing buildings – read in conjunction with A13 – description of the work

- The details of the site boundaries are provided by description and/or by reference to site plans or map references. The estimator will want to know by looking at these drawings or by site inspection, the road network for ease or restrictions of access, which may affect the hours of work, ability to unload from the main road and/or times of delivery. The drawings show what land, if any, is not being built on allowing space for the provision of site accommodation (A41) and material storage. If the footprint of the proposed building takes up most of the site then the public will have to be protected against the possibility of falling objects and perhaps footpaths diverted for safety reasons, all of which can be quite expensive. These drawings would normally detail existing trees in terms of species and size, which if being left in place, may have to be protected and could affect the building work.

- Details of existing buildings adjacent to or on the site. This could have an effect on the methodology and costs of work due to their proximity to the new construction. Examples include limitations in tower cranes over-flying other buildings (this is considered as trespass unless permission is sought and gained) or sensitive buildings such as historical ones, which may be easily damaged or limit access. It is important to note that the contractor will be well advised to document the state of adjacent buildings once the contract is awarded, to avert claims for damage at a later stage. The use of these buildings is also important. For example, if the adjoining building is a dairy then the production of dust from the construction process would be problematic (A34). If the site is approached from a main road, the road and public footpath will be have to be kept free of mud during the course of the works.

- Existing main services (drainage, gas, electrical, water) on and adjacent to the site should be identified so the contractor is aware of them. This can affect the mode of work and offer the opportunity to connect to existing supplies when executing the construction work. On previously built-on sites, some of the services might not have been located and also may run elsewhere from that indicated. This should be remembered in the context of safety as, for example, an excavator cutting into a live electric main could have disastrous results.

- Any other information that is available about the site, which may assist the estimator. This may include soil surveys and likely contamination implications. The detail of the soil surveys is important as the number and position of trial bore holes drilled over the area of the site determines the accuracy of the overall site soil survey as it is easy to miss pockets of poor soil such as peat, if inadequate or inappropriate sampling has taken place. As an example, a square plan building was designed for a site on which an adequate number of trial bore holes were taken for its purpose, but subsequently it was decided to change the orientation of the building by rotating it by 45°. Some of the trial bore holes were now outside the footprint of the building.
- Contamination can be identified from the soil samples taken, but previous uses of the site can also indicate problems, for example if a garage existed then there is a likelihood of petroleum spillage over a long period of time that would have polluted the sub-soils.

A13 – Description of the work

- To be read in conjunction with A12, this is a general description of the work to be carried out, giving for each building to be constructed/modified/refurbished its main dimensions and shape. This would include the plan area and perimeter at each floor level, heights between floors and the total height of the building. It would advise of details of work being carried out by others that may impinge on the contract and identify any other unusual features or conditions.

9.5.2 Contractual matters

A20 – The contract/sub-contract

These are details of the form of contract to be used in the event of a successful bid. A typical example would be the JCT05. The estimator will be especially interested in special conditions such as:

- How much retention is to be made at the end of the contract and for how long will it be held?
- The employer's insurance responsibility, any amendments made to the standard conditions, any performance bonds required and whether or not the contract is to be executed under hand or under seal. Under

hand means the contractor's liability for most defects is for a period of 6 years after completion, and under seal for 12 years. The type of contract used will depend on the method of procurement (Chapter 8). Contract is a significant and complex subject in its own right and is not developed in this text. It should be noted the JCT forms of contracts were designed to be fair to both signatories to the contract. Any onerous additions or removals of clauses, by definition, will favour one or the other.

- If the contract allows for fluctuations, then at what dates do these come into effect?
- Any warranties that may be required for design work carried out by a sub-contractor. There are separate forms for this named Collateral Warranty Form of Agreement.

9.5.3 Employer's requirements

A30 – Employer's requirements: tendering/sub-letting/supply

- There may be further conditions laid down here not in the original invitation to tender letter. These may involve extra costs to the contactor for providing, for example, a guarantee bond which covers the employer in the event of the need to employ another contractor to finish off uncompleted work up to the value of the bond. This is not used on all contracts, but has been required by local authorities and other public bodies. The bond is usually 10 per cent of the value of the contract, and is taken out with an insurance company.
- These are general conditions laid down concerning the contractual relationship between the contractor and sub-contractors and suppliers. This could include such issues as ethical purchasing policies as well as the general conditions. The purpose is to give the employers the opportunity to ensure that all employed in the supply chain comply with their ethos.

A31 – Employer's requirements: provision, content and use of documents

- Concerned with who has the responsibility to provide drawings, i.e. the architect, consultants, contractor, and sub-contractors. It also defines what should be included so nothing is over-looked, and how they should be used.

A32 – Employer's requirements: management of the works

This may include:

- Stating who has the responsibility for the supervision, co-ordination and administration of all the works. This may well differ depending upon the form of contract and method of procurement.
- Taking out appropriate insurance to cover risks such as fire, vandalism, third-party injuries and damage.
- The provision of a master programme for all the works and ensuring progress is monitored. This may also require advising the employer accordingly at stated intervals. It could be monthly as the employer needs to be aware of any delays or if the contract is ahead of programme, as this can have a knock-on effect on its business, for example the need to bring forward or delay the installation of machinery or shop fit-outs. There will also be financial implications to its revenue stream. The type of programme might also be specified, such as to use critical path analysis and to identify the amount of time allocated for inclement weather.
- Collecting meteorological data on site. The site manager recording the basic information in the site diary often satisfies this requirement. It is important, as not withstanding the comments in the item above, excessive inclement weather could be grounds for an extension to contract. The winter of 1962/63 in the UK was so severe, with sub-zero temperatures for approximately two months, much of the construction industry was brought to a standstill during this period.
- Arranging site meetings at regular intervals.
- Notification of when the work commences and finishes. This is not necessarily the whole contract but could be stages or phases of the work.
- Notifying the employer's agents (the quantity surveyor) before covering up work so that it can be measured correctly. This is most likely, but not always, to occur when working on the foundations and drainage, where work can be very rapidly built over or concealed.
- The keeping of site labour and plant records so when calculating any claims or variations to contract there is factual data.
- Who owns unfixed materials stored on site. Until such a time is reached when construction materials arrive as and when they are required to be built into the building (just in time delivery), there will be significant quantities of building materials stored on the site waiting to be used in the building. Current contract almost invariably means materials delivered each month and not built-in, are measured and paid for in part or total by the employer. Who owns what is an interesting legal

question as if the supplier has not been paid for the material they also have an interest in this.

- Daywork (A55) is a method of valuing work carried out on an architect's instruction when it is not practical to use the rates itemised in the bills of quantities. It is important all parties are aware of the procedures and conditions upon which architects instructions for daywork are issued so there is no confusion.

A33 – Employer's requirements: quality standards/control

These can include:

- The employer needs to be satisfied that the materials and components purchased by both the contractor and sub-contractors meet the specification. This section gives the employer the opportunity to specify which samples of materials and components have to be obtained and what tests need to be completed. The contractor can estimate how much this will be.
- It is sometimes necessary for a sample of work such as a brick wall to be constructed so a standard of workmanship can be agreed. This means that all similar work on site can then be compared against this example of workmanship. Alternatively, a sample of a component(s) be erected on site showing the jointing between elements and fixing to the structure. This is done to show the employer the solution works and what it looks like in practice.
- As a result of the debate on ownership of materials unfixed on site (A32), the employer has an interest in the methods by which unfixed materials are protected and stored. Further it is the contractor's responsibility to replace damaged materials at their expense.
- If defective work or material occurs it is important there is a procedure for sorting out the problem, since ineffective procedures could cause delays to the project. The employer needs to be satisfied the procedures in place will not cause any problems of this nature.
- Also covered in part in A35, the contractor will be expected to clear up and remove all rubbish and surplus material as it accumulates and at the end of the contract or phase of work. This clause might be extended to specify not only that materials have to be disposed of correctly, but also have to be segregated and disposed of in an environmentally friendly way using recognised recycling organisations.
- Before the building is handed over, a detailed check – known as snagging – takes place to ensure all the minor problems are listed and remedial

action taken. Examples of this would be touching up paintwork, ensuring all ironmongery works correctly, doors and windows are fitted properly, and nothing is missing like a screw out of a hinge or a shelf in an airing cupboard. This list can be very extensive depending upon the effectiveness of the quality assurance systems used by the contractor.

A34 – Employer's requirements: security/safety/protection

- Contracting by its very nature is noisy and dirty, but the emission levels can be controlled with careful planning and control. The employer can specify these levels. The contractor can also be limited to the amount of noise made before or after certain times of the day (A35). It may be necessary to make provision to protect rivers, streams and aqua flows from effluent pollution. The Manchester Evening News Arena was constructed very close to the Boddington's Brewery, which obtained its water from aqua flows below where the proposed building was to be built. Any pollution from the construction would have severely affected or indeed destroyed the brewer's business. One should remember that after the contractor has left the client still has to live with the neighbours.
- The public and private roads surrounding the site have to be maintained. This is in part a pollution problem resulting from mud carried off the site by vehicles, but also to do with damage from these and other heavy and tracked machinery.
- Existing and surrounding buildings need to be safeguarded against damage by the main contractor or sub-contractors, and any damage caused made good at their expense and not the employer's.
- The existing underground services will be marked on the site plan. The contractor must ensure these are not damaged in any way from excavation or overloading from either heavy traffic or excess demand. It may be specified the contractor should take all reasonable steps to check these drawings are correct and complete.
- Sites are vulnerable to theft, vandalism, drug users, those sleeping rough, and intruders, which can include children who see the site as a gigantic playground. The provision of security on the site is not just to prevent the problems mentioned, but also to protect the same people against injury to themselves, as there is a duty of care. Employers wish to protect their reputation by not having such incidents on the site.
- Protection to the works as it progresses, especially completed work, from damage by impact, the weather and other trades.
- Protection of existing trees and habitats may be required by law as well as by the client. For example, trees may have preservation orders placed

upon them, and other wildlife including bats, nesting birds and types of newts, as well as certain species of flora, are also protected.

A35 – Employer's requirements: specific limitations on method/sequence /timing/use of site

These are issues which limit the contractor's and sub-contractor's method of operation, their sequence and timing. Examples include:

- Restrictions on the working hours, which may include the start and finishing times, specific days of the week and public holidays. This may be due to the environmental impact to others adjacent to the site or because the client has particular views and beliefs, such as working on the Sabbath.
- Access roads is a term used to describe both the temporary roads on the site and those public roads (A34) used to approach the site. The employer may wish to restrict access from certain roads on the grounds that the road is not substantial enough to take the loads or it goes through say a residential area causing nuisance or a safety risk to children. Temporary roads on a restricted site may be relatively inexpensive but on a complex large site costly depending on ground conditions and distance travelled. This could be also found under A36.
- Any materials which are found on the site and have to be disposed of, excluding material which would be part of a bills rate such as excavate and cart-away or the normal wastage expected in other operations. There may also be environmental considerations stated here.
- Design constraints can be included especially if the contract is such that the contractor is involved in, or has some leeway in selecting materials. The employer may wish to ensure certain choices are made. Any planning restraints may also be identified here.
- The employer could wish to determine the method and sequence of work. This would be particularly important when working in an already occupied building or if the building is to be handed over in phases. It could relate back to items such as noise and pollution as identified in A34.
- It may seem strange on a construction site that a condition be laid down that the use of foul language should be prohibited and if occurring the offender should be instantly dismissed, but the contract awarded for the construction of the Mormon Tabernacle near Preston made that a requirement.

A36 – Employer's requirements: facilities/temporary works/services

- The employer, who may have permanent or semi-permanent representatives on the site, such as members of the design team, the clerk of works and resident engineer, will need accommodation. The contractor will usually be expected to provide these facilities for them and this may include furniture and equipment. This is additional to that required by the contractor (A41).
- This provision could include such items as cleaning and the provision of sanitation, running water, heat, lighting, power and telephone.
- The employer may specify the location on-site of both these offices and the accommodation of the contractor.
- Temporary hoardings around the site are often used by the employer to advertise themselves and they can lay down some strict conditions accordingly in terms of colour, quality and specification of material used and so on. If not, the contractors may paint them in their own livery colours and advertise themselves.
- Normally the employer, design team, contractor and sub-contractors advertise themselves using name boards. There may be limitations in the overall size and position of the name boards laid down by the local authority planners or by the employer. It is not unusual for the dimensions of the individual boards to be controlled and specified and in some cases the employer may only want their name to be shown because of conflict of interests. For example, if the contractor is also involved in property development, the employer may not wish this fact to be advertised.

A37 – Employer's requirements: operation/maintenance of the finished building

- This is the provision of facilities and services to be provided by the contractor and sub-contractors on completion and/or after, to assist the employer operate and maintain the finished building. This could include instruction and maintenance manuals, training for maintenance staff especially on the mechanical and electrical side.

9.5.4 Contractor's general cost items

Further information can be found in *Operations Management for Construction*, Chapter 1.

Table 9.6   Site staff

Contracts manager	Planner
Site manager	Assistant planner
General foreman	Quantity surveyor
Trades foremen	Assistant quantity surveyor
Ganger	Safety officer
Engineer	Secretarial staff
Assistant engineer	Security staff
Chainman	Cleaners

A40 – Contractor's general cost items: management and staff

The contractor's management team will vary considerably depending upon the size and complexity of the project. Most contractors will have a standard list of personnel likely to be used, which they can use as a checklist. It should be noted that the cost of staff is not just their salary, but includes:

- pension scheme (employers contribution)
- annual bonus
- overtime
- car and expenses
- training levy.

The amount paid out will be different each month, as the staff required to manage the contract will change as the contract progresses. Typically the type of staff will include, on a medium to large contract, those shown in Table 9.6.

A41 – Contractor's general cost items: site accommodation

The type of temporary accommodation required by the contractor is dependant upon the needs in managing the contract itself and is in part determined by the numbers of staff decided in A40. What is provided will also depend upon the space available to site the accommodation. Note that this does not include for any accommodation provided for the employer (A36). On inner-city development the accommodation may also be off site, with the contractor renting other premises. Typical accommodation may include:

- offices for the management and administrative staff;
- stores for small and/or valuable components, materials and hand tools;

- canteen used for lunch breaks and a place to go during inclement weather;
- toilets, both male and female;
- drying room to hang wet clothes overnight;
- first-aid facilities on a large site, if a full-time first aider/nurse is employed;
- washing facilities;
- security cabins at the entrance and exit gates of the site.

The total costs of providing these include erecting and fitting-out, removal after the contract is completed, transport to and from the site, foundations and drainage, furniture, rates and insurance.

A42 – Contractor's general cost items: services and facilities

This comprises many items. An indicative list with comments is given here, but is not necessarily complete. Some of these will also have to be provided as attendance to sub-contractors.

- Power, lighting and heating: the availability of supply needs to be ascertained. Is it sufficient to run a tower crane and how far is the connection away from the site? The cost includes connection and disconnection charges as well as the running costs.
- Water: similar issues as with the electric supply, but if not readily available the use of bowsers can be considered.
- Telephones: the number of lines required for telephony and IT needs to be established, as this will indicate the costs of connection and the monthly costs. Note the the costs of these provisions (electricity, water and telephone) for the employers (A36) could be incorporated here, when the estimator is pricing the bills of quantities.
- Stationery and postage for managing and administering the project.
- Office equipment (as distinct from furniture): which today would likely comprise facsimile facilities, photocopying and computers. There will also be the added burden of security costs.
- Health and safety: the contractor will be expected to provide employees certain protective clothing including safety helmets and goggles, safety equipment such as harnesses, barrier creams, signing around the site and training. It will also be necessary to provide appropriate protective clothing for visitors and often training before allowing them on site.
- Storage of materials: this can involve secure compounds, racking, etc., to support and protect materials, tarpaulins, and preparing the ground.

- Provision of temporary electricity for plant and site lighting.
- Security: this has become a very important issue, not just because of potential theft and vandalism, but also the risk of terrorist activity. The latter depending much on the location and/or purpose of the building such as an airport terminal, government or prestigious building. There will be security during the working day and outside working hours. In the case of the former, checking in and out procedures to ensure only bona fide people are on site and in the latter patrolling the site. It will also mean providing material security items such as locks and alarms.
- The cost of providing the materials and labour for the protection of work in progress (A34).
- Cleaning of work in progress and of the whole building before handover to the client.
- Waste management: providing skips and waste chutes to collect and, increasingly, to permit the segregation of waste to enable more of the waste materials to be recycled rather than sent to land fill tips.
- Small plant and tools is sometimes taken account of here, but alternatively can be costed with mechanical plant (A43).

A43 – Contractor's general cost items: mechanical plant

There are differing views whether or not small tools should be here or in A42. It makes no difference in reality providing they are costed somewhere. Small tools and plant could be a very long list, but Table 9.7 gives an idea.

Table 9.8 shows some large items of mechanical plant.

Some of these items may be included in the all-in unit bills rate (section 10.11) and would not be included here. This would be the case if

Table 9.7 Small plant and tools

Barrows	Saws
Blocks and tackle	Water pumps
Steps	Concrete vibrators
Trestles	Concrete screeders
Buckets	Compressors
Pick axes	Generators
Spades and shovels	Fuel tanks
Drills	

Table 9.8 Mechanical plant

Excavators and earth movers	Concrete mixers and silos
Concrete mixers and pumps	Lorries
Dumpers	Hoists
Forklifts	Piling plant
Mobile cranes	Pavers
Tower cranes	Tractor and trailers
Compressors	

a piece of plant is being brought in for a specific, relatively short defined operation, as distinct from a dumper say, which may be on the site for the majority of its construction programme and used for a multitude of different bill items. Other equipment, which is being hired, may also include for the driver's time if this is provided as part of the hire agreement such as with large excavation plant. The running and maintenance costs will be provided either by the contractor and/or the hirer.

A44 – Contractor's general cost items: temporary works

Some of these items may have been taken account of in A34 and A36 as employer's requirements, but will probably be priced by the estimator here.

- Besides satisfying the employer, the contractor is also interested in access roads on and onto the site. Approach routes to the site are important to ensure goods coming to the site have unobstructed access and to establish the best routes in and out of the site to effectively move goods (*Operations Management for Construction*, Chapter 1).
- Strengthening existing roads could have already been specified (A34), but if the road is not substantial enough, then a cost will be incurred to carry out the necessary works, as it is the contractor's responsibility to protect the highway.
- Temporary walkways. It is sometimes necessary to divert public rights of way running through the site as well as re-routing public pavements adjacent to the site.
- Access scaffolding is required on the outside of most building projects for access, platforms, towers and inside for stair- and lift-wells. It will also be used as guard rails to prevent people falling into holes, from the

building or into a trench. This will also include scaffolding boards and toe boards used for walkways and for ladders for vertical access.

- Support scaffolding for formwork to stairs and soffits and propping once the formwork has been removed until the concrete has reached its design strength.
- The contractor requires hoardings and fencing for the same reasons as outlined in A36. They are also required as part of the security strategy for protection of the site.
- If traffic from the site has to cross a highway or if the works partially obstructs the flow of traffic it may be necessary to include for the provision of some form of traffic control. This can be achieved either by automated lights or using labour.
- The topic of positioning and using notices and signs to advertise is discussed in A36. However, the other purpose of showing the company's name is to direct and assist visitors and those delivering materials to locate the site. This can mean the provision of 'named arrows' on street lamps some distance away from the site, although this does need permission.
- Any other temporary structures required such as bailey bridges are priced here.

9.5.6 Work carried out by others

A50 – Work/materials by the employer

This is works carried out, or materials supplied, by the employer. The cost to the contractor is for any attendance required and is based on the same criteria as for nominated sub-contractors.

A51 – Nominated sub-contractors

All work carried out by nominated suppliers has to be entered and described in the Preliminaries. A nominated supplier is one that has been selected by or on behalf of the employer to carry out a specific part of the work. This section determines how they should be covered in the bills of quantities. Ideally the information contained here should be reasonably comprehensive and include a list and cost of the main elements of work, the time required to complete the work and the location of any plant that might have to be used. This gives the contractor the information to programme the works, which can have an effect on the overall programme and costs. The contractor is permitted a percentage attendance allowance to support them plus a percentage profit. It is possible for the contractor to refuse to use them, but there can be problems associated with this. The nominated sub-contractor's price shown here is known as a prime cost sum (PC sum).

A52 – Nominated suppliers

As with the nominated sub-contractors a PC sum is entered here allowing the contractor to add a percentage profit.

A53 – Work by statutory authorities

This is the work to be carried out by any or all of the statutory authorities – water, gas, electric – and is entered as a PC sum.

A54 – Provisional work

A provisional sum is put in the bills of quantities for items that cannot be described or itemised. There are two kinds – defined and undefined – both for work that has not been completely designed.

- defined – the nature and quantity can be identified and the contractor has to allow for a place in the programme and also in pricing the preliminaries
- undefined – the work in not yet defined so the contractor will be paid for doing the work, planning it and is given reasonable overheads for it.

A55 – Dayworks

This is a way of valuing work carried out as an architect's instruction after the works has commenced and when there are either no equivalent bill items, or the quantity is such that it would be unfair to use the rate in the bill. For example, it is very different, from a productivity point of view, to pour $50m^3$ of concrete as compared with $2m^3$. Cutting holes through slabs and walls are examples where there may not be an appropriate bill item. The work done is recorded in terms of the labour, material and plant used and agreed by the employer's site representative, usually the clerk of works, but it could also be the architect.

9.6 The bills of quantities

9.6.1 Introduction

The main purposes of the bills of quantities are as follows:

- It enables all the contractors tendering for the contract to price on exactly the same information in a form that is standard (SMM7).
- The risk to the contractor is minimised as each part of the work is clearly itemised, theoretically resulting in more realistic and competitive tenders. The proviso to this is that the completed bills total price may not be the same as the employer finally has to pay because of subsequent variations and claims.
- If a more accurate price is required then the employer and the design team will be encouraged to finalise the design and have produced a full set of drawings and specifications upon which the bills can be based.
- After the contract has been let, it acts as a satisfactory basis for the calculation of variations made as a result of design changes occuring during the contract period.
- It acts as a good vehicle to calculate the certified payments throughout the contract. (Certified payments are for the work done each month and any materials delivered during the same period, but not yet incorporated into the building.)
- As it is a quantified and described itemised list of all components and materials to be used in the building, it gives the contractor the opportunity to negotiate and place provisional indicative orders of the bulk materials and components with suppliers.
- The same data can be used by the contractor's planner for programming the work and calculating the labour resource required.
- Can be used as a cost-control system by the contractor to ensure the work is within budget.
- It is the basis for calculating and agreeing the final account after the completion of the contract.

9.6.2 Producing bills of quantities

This section gives an overview, as details of bills of quantities can be found in the given references on this subject, but some understanding of the procedures to realise why the process is lengthy is appropriate. As indicated earlier the basis for the bills of quantities is the Standard Method of Measurement that

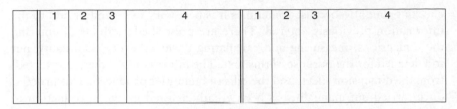

Figure 9.1   Dimension sheet

defines each item of work in a standard and agreed manner. The first part are the Preliminaries detailed in section 9.4. The procedure of the remaining items follows a similar pattern to that described below. The first stage is to take off all the quantities from the drawing, carefully describing and measuring them as defined by the SMM.

This is done using dimension paper. The layout of dimension paper conforms to the requirements of BS3327: 1970 Stationery for Quantity Surveying and is produced in A4 paper size format, a part of which is shown in Figure 9.1 (not to scale).

The dimension sheets are divided into two identical ruled parts each comprised of four columns that are used as follows and are described in their order of use:

4. The 'description column' in which the written description of each item is entered as defined in the SMM7. Sometimes the right-hand side of this column is used for preliminary calculations such as working out the length of the centre line of the perimeter of a wall. These are sometimes referred to as 'sidecasts' and are important to show as if there is any query about the validity of this taking off process, the calculations can be easily traced and checked.
2. The 'dimension column' in which the actual dimensions measured from the drawings are entered. This can be one, two or three dimensions, depending on whether or not the item being measured is linear, square or cubic (m, m², m³) or items such as one number door.
1. The 'timesing column' is used for multiplying when there is more than one item being measured that is identical. For example the concrete in a strip foundation of a house may have been measured, but there are 10 other identical houses on the estate.
3. The 'squaring column' is where the total length, area or volume is calculated by multiplying the figures in columns 1 and 2 together.

Those carrying out this operation are advised to allow plenty of space and not to cramp the entries on to the dimension paper partially for the ease of checking, but also because sometimes it is necessary to amend or add to the information previously entered. There are prescribed methods of entering the dimensions, measuring and calculating covering every eventuality, but too detailed for the purpose of this text. The information is then 'abstracted' from the dimension sheets and the bills of quantities produced. This process, once carried out manually, has been greatly assisted by the introduction of modern software.

References

Seeley, I.H. and Winfield, R. (1999) *Building Quantities Explained*, 5th edn, Macmillan.

Standard Method of Measurement of Building Works (1997), 7th edn, RICS.

Willis, C.J. and Newman, D. (1995) *Elements of Quantity Surveying*, 8th edn, Blackwell Science.

10

Estimating and tendering procedures

10.1 Introduction

This chapter is concerned primarily with the invitation to tender using the traditional method of procurement, although much of what is contained here is also relevant to other methods. The term client is preferred over employer by the author as it more properly reflects the relationship with the contractor and reduces confusion when refereeing to employer/ employee relationships in an organisation. It is interesting that in Total Quality Management the term 'customer' is used (*Operations Management for Construction*, Chapter 8).

It is important to note the estimator's prime role is to estimate, as accurately as possible, the true cost of doing the work on the site. This excludes any overheads and profit management may wish to add. This net cost is called the estimate. The tender price is that which is quoted to the client (the employer) and is the estimate plus the overheads and profit. If the estimator has a proven record of being accurate with the estimate, the more confidence management has in making strategic decisions as to the amount of overheads and profit they should add to be truly competitive, and being surer of attaining their planned financial returns for the business.

The more comprehensive and accurate the documentation provided by the client, the more likely the estimate will be close to actual, and equally important, the likelihood of future claims, disputes, variations and delays in completing the contract will be reduced. This is very important as it is negotiations over these financial claims that can lead to confrontation rather

Table 10.1   Causes of problems with information

Cause	Reason
Missing information	It has not been produced, or has but has not been sent to the contractor
Late information	Has not arrived in time for the contractor to plan the work properly or is too late for materials to be ordered and delivered on time
Wrong information	The information is out of date, or includes errors such as dimensions and descriptions
Insufficient detail	Inadequate information for either tendering or construction processes
Impractical designs	The design solution is either difficult to construct or will not work in practice
Inappropriate information	Either not suitable for the purpose intended or not relevant
Unclear information	Caused by poor draftsmanship, or the solution is ambiguous
Not firm	Provisional information which is indistinguishable from firm information
Poorly arranged information	Not produced in a consistent way, poor titling and generally difficult to read as a result
Uncoordinated information	Where it is difficult to read or relate one drawing with another
Conflicting information	Where documents disagree with each other. This often occurs between the architect's and the engineer's drawings

than co-operation between parties, as proposed by both Latham (1994) and Egan (1998).

Unfortunately, the documentation provided at the tender and construction stages is not always accurate. Table 10.1 (adapted from the findings of the Co-ordination Committee for Project Information, 1987) identifies typical examples of deficiencies in the provision of information for the tendering and construction processes and the reasons why. However, it should be noted that many of these are more likely to happen at the construction stage.

10.2 Selection and invitation to tender

Invitation to tender via the traditional or design and build methods of procurement, comes about in primarily three ways:

1. The client or representatives advertise in the press or trade papers and ask for contractors to tender. The contractor wishing to tender for the work, replies to the advertisement and in return receives the contract documents. Sometimes a deposit is required, refundable on receipt of the completed bid so as to discourage frivolous enquiries. This process is referred to as 'open tendering'.

   Public contracts over 5 million euros in value must be advertised across Europe in line with a European Directive so free trade can occur across the European Community. These can be found in the *Official Journal of the European Union (OJEU)*.

2. More often used is 'selective tendering', when a list of perhaps five to eight appropriate contractors is drawn up and they are asked to tender. This list is referred to as a preferred list.

3. In this final case only one contractor is approached because of its track record in the particular type of work and the design team wish to avail themselves of the contractor's knowledge during the design stage. It is highly probable that the client and/or design team has worked with this contactor in the past. This is known as a 'negotiated tender'.

In selective and negotiated tendering the experienced client will select contractors to tender based on the following:

- The size of the company may determine the range of work they can be expected to manage successfully. Large contractors often have difficulty managing small contracts cost-effectively as the overheads may be too high. Similarly, a small contractor may not have the ability to manage a very large contract. The quality of the management team is measured on how successful they are at maintaining relationships with labour sub-contractors, and completing similar work well and on time.
- The experience the client has had with the contractor on previous work would reinforce the above knowledge.
- The contractor's financial status is important. Indicators, although not always reliable, would be the length of time the company has been in business, financial checks carried out, bank references provided and insurance cover.
- Finally, the client may wish to know the contractor's current workload. If over-stretched, another tender may be too much or it may not be able to devote enough resources to the project.

Clients do not always choose the lowest price tender. A client may prefer the security of a company it knows, or it is believed the lowest bid is too low and could put the winner in danger of going bankrupt.

10.3 Tender documentation

When the tender documentation arrives, the estimator will check that all the information stated in the letter of invitation is enclosed, there is sufficient information to permit a realistic cost of the contract to be produced and adequate time has been allowed to produce an informed and accurate tender.

Typical sources of information that come with the letter of invitation are:

- Drawings that will be at least those used to produce the bills of quantities and may include others done since. There should be site layout drawings giving information about access roads, boundaries, topography, adjacent buildings and structures, services on the site, roadways, trees, restricted areas and the footprint of the proposed building. General arrangements such as plans, sections and elevations are needed as are any specialist drawings done such as temporary works and structural surveys. Detailed drawings should include substructure, frame, floors, roof, cladding, internal structure, internal finishes, and services.
- Schedules of information including standard and non-standard joinery details, windows, doors, ironmongery, sanitary fittings, partitions, and floor and ceiling finishes.
- Descriptions of performance specifications, if appropriate.
- Site investigation reports including soil surveys, water table levels, and other technical reports such as the construction (design and management), health and safety plan, reports on existing asbestos and contaminated ground.
- Bills of quantities.
- The form of contract such as the JCT05 and any supplementary conditions.
- Programmes of work periods for the main nominated sub-contractors if not in the contract form.

This information can and should be supplemented by carrying out a site inspection to understand the likely production problems due to space and traffic restrictions, the state of existing or adjacent buildings, neighbours and effects on them from noise, dust and dirt, ease of access, ground conditions, over-flying of tower cranes and other issues already highlighted in the employer's requirements and contractor's general cost items in the SMM7 (section 9.5).

Other issues would include alternative solutions if there was insufficient space on-site for staff accommodation, the location of the nearest landfill site and the charges, the local transport system, parking facilities and their costs, which can have implications in recruiting labour. Also, what are the local facilities for food and accommodation? Is the site exposed to extreme weather conditions or the possibility of flooding?

The statutory authorities will be contacted regarding supply and on controversial contracts it will be necessary to assess the likely disruption by protestors.

10.4 The decision to tender

On receipt of an invitation to tender, the contractor has to decide whether to accept or decline the invitation. If declined, the documents are returned to give another contractor the opportunity to tender. The estimating procedure using full bills of quantity is very time-consuming and costly. There is no point in expending this effort unnecessarily, so when an invitation arrives the contractor will consider the following, set against the background of the company's current workload:

*   Is the size of the contract too small or too large? If too small, the culture of the organisation may be inappropriate to manage the works, and overheads will make the bid uncompetitive. If too large then the company's resources may not be able to cope. The first M1 motorway project was divided into four separate contracts. John Laing bid for all four and although not the cheapest on each, the client asked them to consider constructing all four as one contract. The total value of the works was very high relative to their annual turnover. Laing's board made a strategic decision to take on the work.
*   Has the company experience of this type of work? Again, this may become a strategic decision to break into a new market. Equally, it could be removed from the company's experience and the decision made to decline.
*   Even though the company has experience in the type of work, have they the experience in the type of contract? For example the JCT05 used for traditional procurement as against design and build.
*   How much competition is there? It is said that the success rate is about one in four to six bids and these odds can be built into the overhead

calculations. If there is a lot of competition then the odds will lengthen, confusing these calculations.

- Is the proposed contract in the company's usual geographic boundaries? If not, then tendering becomes a strategic decision.
- Does the company have the necessary resources, especially manpower, to cope? If not, the company will need to recruit the necessary staff. The implications of this can be considerable as it takes time to recruit and train suitable staff and, assuming they are available, will they expect continuous employment? This means more work would have to be found, which means the company is looking at an expansion policy and all that entails.
- Has the estimating department the capacity and expertise to tender? The latter is important if the work is different to that normally tendered for.
- Are the documents provided comprehensive? If not then, the estimating department will have more difficulty in reaching a true cost. It also raises the alarm as to how the contract might be serviced by the design team during the construction phase and increases the likelihood of claims and confrontation.
- Has enough time been allowed to prepare an accurate tender? This can affect accuracy if insufficient time cannot be allocated to fully comprehend all the implications of constructing the project and could increase the likelihood of claims and compensation.
- What is the current workload of the contractor at the time? If the workload needed to meet the annual plan has been achieved, and management decides to tender, it can add lower than normal overheads to become more competitive. This is slightly more complicated than just how much work has been successfully bid for, as it must also take into account the chances of it being obtained, when it commences and over what duration.
- What is the financial status of the client and what is its track record for settling accounts?
- Supplementary contract conditions attached will be insurances, warranties and performance bonds that are quite straightforward to price and common to all bidders. However some non-standard conditions can be onerous and the contractor may wish to decline from tendering, re-negotiate these conditions or add a premium.

Two issues occur regularly: strategic decisions and risk. In both cases senior management will be involved. (Strategic thinking is developed further in *Business Organisation for Construction*.) However, management may have statistical information about their previous performances against the current competitors (section 11.2), and this can be brought into the decision-making process. Risk can be managed with the knowledge that some of the burden of risk can be moved from the client to the contractor and that uncertainty increases the risk. In both cases the contractor needs to consider how much, if any, should be added to the tender price to take account of this increased risk.

10.5 Managing the estimating process

Once the decision has been made to tender for the project, the estimating department needs to control the work being processed through the section. To do this the company will usually keep a tender register that gives a summary of the status of all contracts. This can be used to help manage tenders, and as record of past successes and failures in case any obvious trends have occurred. Typical headings are shown in Table 10.2.

The client column might be extended to name the architect, structural engineer and quantity surveyor (QS) used by the client. The percentages, low and high, in the results column are established by the client advising the contractor what the other bids were, or through second-hand knowldege. Irrespective of the source, this indicates the current market value as perceived by one's competitors and it acts as a benchmark for future bids. Although not shown in this table, the contractor should keep a record of competitors' bids for each contract to assist in developing a bidding strategy (Chapter 11).

Many estimators relate that there is rarely enough time to complete the estimating process and that as the final submission date nears, the department 'panics'. Therefore it is important that all estimates being processed through the department are programmed and monitored, as this demonstrates the overall workload. Each individual contract has its own programme. On the

Table 10.2 Sample tender register

Tender register	Period from:						
Tender reference no.	Client	Contract name	Tender		Result		Comments
			Submission date	Value £s	Yes % low	No % high	

programme not only should each item of work be shown, but also which person in the department should be responsible for executing the task. Typical tasks that should be identified and entered on the programme, some as action dates and others as activities are described below and summarised in Figure 10.1.

- The date the documentation was received from the client.
- Select the estimator responsible to run the project. This might be after the decision to proceed, as the chief estimator may wish to check if the documents to provide enough information to decide whether or not to tender. This occurs shortly after the receipt of the drawings and must give enough time for an informed decision to be made, and leave sufficient time, if the answer is yes, to permit the tendering process to be accomplished properly.
- If not already done a site visit is organised as are visits to consultants to inspect other available information.
- With all of the data available the methodology for construction can be determined and a pre-tender programme produced (*Operations Management for Construction*, Chapter 2). Much of the work contained in the bills will be sub-contracted to others, so after analysis of the bills, sub-contractors are selected and invited to tender. Equally prices for materials not to be included in the sub-contractor's quotation have to be obtained. Approximate dates and durations can be abstracted from the programme enabling realistic quotations to be submitted (section 10.15).
- A deadline by which all enquires must be sent out.
- A deadline for receipt of all prices of materials and sub-contractors. The former is usually required earlier as this affects the pricing of items in the bills especially if the contractor is carrying out the work itself.
- The pricing of the bills can commence whilst awaiting the return of this information, especially many of the 'preliminaries' items.
- The most competitive and realistic bid from material suppliers and sub-contractors is selected and then entered in the bills.
- A deadline for completion of the pricing of the bills is established.
- On completion of the priced bills it is important to check the calculations.
- Cash-flow predictions can now be calculated.

TENDER TIMETABLE

Description	Key date	February																				March						
		1	2	3	4	5	8	9	10	11	12	15	16	17	18	19	22	23	24	25	26	1	2	3	4	5	8	9
Documents received	1 Feb	■																										
Select staff	1 Feb	■																										
Inspect documentation			■																									
Decision to tender				■																								
Site visit						■																						
Visit consultants							■																					
Determine methodology								■																				
Pretender programme	5 Feb								■	■	■	■	■	■	■	■	■											
Abstract, despatch subs.	10 Feb					■																						
Abstract, despatch mats.						■			■																				
Receive quotations mats.	24 Feb																		■									
Receive quotations subs.	1 Mar																				■	■						
Price labour and plant										■	■	■	■	■	■	■	■	■	■									
Price materials																						■	■					
Price sub-contractors																							■	■	■			
Price preliminaries																								■	■	■		
Add overheads																									■	■		
Check																												■
Review meeting	10 Mar																											
Submit documents	11 Mar																											
Submission date	12 Mar																											

Figure 10.1 Tender programme

- A date is fixed for a meeting of the senior management to determine the overheads and profits.
- Time allowed for preparing documents for submission and on the date stated by the client.

10.6　Completing the bills

Many of the activities listed in section 10.5 run concurrently, but for ease of understanding they are discussed separately. The estimator's tender programme (Figure 10.1) shows these overlaps. Whilst the preliminaries items come first in the SMM7, the decisions on whether items such as plant should be included in the bill item or the preliminaries needs to be determined.

10.7　Building up the costs of an item

In section 4.6 under General Rules in the SMM7 it states, 'In a bill of quantities, the following shall be deemed to be included with all items:

Labour and all costs in connection therewith.
Materials, goods and all costs in connection therewith.
Assembling, fitting and fixing materials and goods in position.
Plant and all costs in connection therewith.
Waste of materials.
Square cutting.
Establishment charges, overheads and profit.'

The establishment charges, overheads and profit will be excluded from the build up of the unit rate as these can vary depending on the management decisions at the time of tender (section 10.22). The other items are labour, materials and plant costs.

Each item in the bills of quantities is based upon the SMM7. The process of calculating rates does not occur every time a new tender arrives for pricing. A comprehensive library of rates based on these calculations will have been developed over the years. However, changes will need to occur from time to time as costs increase, so they will be regularly updated. Occasionally, new rates have to be produced in the case of some special requirement not covered in the database.

10.8 Calculating labour costs

The costs of labour to the employer are not just the basic rate, but include many other items:

- The standard wage which is the basic hourly rate multiplied by the weekly working hours agreed and published in the National Working Rule Agreement (NWRA).
- Overtime, which can be the hourly rate multiplied by 1.5 (time and a half) or 2 (double time).
- Guaranteed minimum bonus which is laid down in the NWRA as a means of ensuring a bonus is paid if the operatives are unable to work as a result of, say, inclement weather.
- Bonus that is calculated based on output (*Business Organisation for Construction*, Chapter 8).
- Plus rate, either paid in lieu of bonus or to attract labour if in short supply. This is an hourly rate in addition to the basic wage rate.
- Extra payments under the NWRA for special skills such as driving and maintaining mechanical plant.
- Travelling costs that can include the time it takes to get to site as well as fares. It must be remembered that the site may be a long way from where the employee was engaged.
- Lodging allowances may have to paid.
- Construction Industry Training Board (CITB) levy paid as a percentage of all employees' wages to fund training for the industry as a whole (*Business Organisation for Construction*, Chapter 8).
- Public holiday pay.
- Other holidays which operatives can be entitled to in their conditions of employment.
- Employer's contribution to pensions, accidental injury and death cover schemes.
- Sick pay, which has to be paid up to a certain point, when operatives are ill or injured.
- National Insurance contributions by the employer.
- Severance pay if the employee is dismissed prematurely.
- Employer's liability and third-party insurance.
- Tradesman supervision. The trade foremen may be employed as a supervisor full time or part time, the remainder of the time working on their trade.

From all this information an hourly rate can be calculated. This calculation needs to be done for both craft and general operatives because their basic wage rates are different. Adjustments can be made to the calculated rate to take into account:

- Overtime working may well vary depending upon the time of year. For example, it is more likely to be worked in the summer months when the daylight hours are longest.
- The amount of skilled labour available in the area which can affect supply and demand or the need to bring operatives in from another area.
- What percentage of operatives are to be paid travelling expenses and how long does it take to travel in the particular area.
- The part of the country/world and the effect of climate.
- The part of the world and likely productivity and methods of work.

Organisations will do their calculation in slightly different ways, but the calculation shown is a typical method that is applied and has been adapted, with permission from the Chartered Institute of Building, *Code of Estimating Practice Endorsed by the CIOB*, 6th Edition (1997).

There are basically three components in the calculation: to establish the hours likely to be worked over the year; to calculate the annual wages; and to establish the average hourly rate to be used in the bills rate.

10.8.1 Average annual working hours available

Before reading on, the author wishes to issue a health warning about the figures used in this section. The hours of work, conditions of employment and the costs to the employer all change regularly as a result of legislation and nationally negotiated wage agreements. The figures, therefore, should only used as a means of demonstrating the principles and not as actual figures.

An assumption is made that the basic week is 37.5 hours worked throughout the year, comprising 7.5 hours per day Monday to Friday. As the hours of daylight vary throughout the year, in the northern hemisphere summer provides more opportunity to work in natural light than winter. It would not be unreasonable to assume there are 30 weeks available for overtime work in a given year. Not withstanding EC legislation on working hours, employers may negotiate special working conditions and hours of work on a specific contract because of the nature of the work, such as working shifts.

The employer pays the employees for public holidays and annual leave. If 21 days of annual leave are paid for, it could be assumed that 7 days are taken at Christmas, 4 at Easter and 10 for a vacation. The latter two could occur during the summer period when extra overtime would be worked. In the UK there are three public holidays during the winter period – Christmas,

Boxing Day and New Year's Day – and five in the summer: Good Friday, Easter Monday, Early May Bank Holiday, Spring Bank Holiday, and Summer Bank Holiday. The calculation for the total weekly hours worked is shown in Table 10.3 and the annual hours in Table 10.4.

The total hours operatives are available to work in a year are 750 + 1179 = 1929. On average, operatives are usually off sick in the winter for 8 days, a total of 60 hours giving a net hours available for work of 1929 – 60 = 1869.

A further deduction should be made for inclement weather, as it is almost certain that work will stop at some time during the year as a result of excessive rain, snow, winds or sub-zero temperatures. The amount deducted will depend on the level of exposure of the site, its geographic position and time of year. Assumptions made on this may well change in the light of climate change. It is not unusual to assume a loss of 2 per cent, which in this case would mean losing a further 37 hours, making a figure of hours actually available for work as 1869 – 37 = 1832.

Table 10.3   Weekly hours worked

Winter period		Summer period	
Starting time	8.00am	Starting time	8.00am
Lunch	1.00–1.30pm	Lunch	1.00–1.30pm
Finishing time Mon–Fri	4.00pm	Finishing time Mon–Fri	5.30pm
Total hours	37.5	Total hours	45

Table 10.4   Annual hours of work

Winter period		Summer period	
Working week	37.5	Working week	45.0
For 22 weeks	825.0	For 30 weeks	1350.0
Deduct		Deduct	
7 days leave*	52.5	14 days leave**	126.0
3 days public holiday*	22.5	5 days public holiday**	45.0
Total hours	750.0	Total hours	1179.0

* A 7.5-hour day, ** A 9-hour day

10.8.2 Calculation of the total wage costs

Having arrived at a figure for the estimated annual number of hours operatives are available for work, i.e. 1832, the actual annual cost can be calculated. This means taking the basic hourly rate and adding on all the other costs incurred by the employer in employing staff.

Basic wage

For the purposes of this calculation a figure for the wages has been assumed to be:

General operative: rate £240 for 37.5-hour week or £6.41 per hour
Tradesman: rate £298 for 37.5-hour week or £7.95 per hour

Therefore, the basic annual wages excluding overtime rates are:

General operative: 1869 hours × £6.41 = £11,980.29
Tradesman: 1869 hours × £7.95 = £14,858.55

Note this includes the 37 hours for inclement weather, as the operatives have to be paid even though they are not working.

Overtime

Overtime is paid for any time worked in excess of the basic working week (37.5 hours). In the calculation shown in section 10.8.1 no overtime is worked during the winter period, but a prediction of 7.5 hours per week is made for the summer period. This means that operatives working this amount of overtime are paid at time and a half. In other words they will receive 11.25 hours pay for the 7.5 hours worked. Whilst the operatives earn an extra 3.75 hours pay in overtime, the contractor still only obtains 7.5 hours output so the overtime paid is considered non-productive time. The total cost of this non-productive time over the complete summer period is:

30 weeks less 4 weeks leave and public holidays × 3.75 hours = 97.5 hours

The cost of this is:

General operative: 97.5 hours × £6.41 = £624.98
Tradesman: 97.5 hours × £7.95 = £775.13

Bonus payments

Operatives will also earn bonus at, say, one-third of the basic weeks wage, which is £1.93 for general operatives and £2.39 for tradesmen. This works out, for the 37.5 hour week, as:

General operatives: 37.5 hours × £1.93 = £72.38
Tradesman: 37.5 hours × £2.39 = £89.63

Sick pay

It was assumed that on average 8 days sick leave is taken on by each operative and this would occur in the winter period. No payment is made for the first three days (known as qualifying days) so the employer, in this case will only have to pay for 5 days. The Department for Work and Pensions determines the amount of payment, which is on a sliding scale depending on income. At the time of writing the basic sick pay is £12.87 per day and this is used for this calculation.

Trade supervision

Contractors and sub-contractors will employer trade supervisors or foremen. The number of operatives they are responsible for will vary from company to company and the complexity of the works. Often these appointments are such that they spend half their time supervising and the remainder engaged in their trade. For the purposes of this exercise it is assumed they are responsible for six tradesmen, are supervising for 50 per cent of their time, and are paid an extra 50p per hour for the added responsibility.

The hourly cost for the gang is:

1 trades foreman £7.95 + £0.50   = £   8.45
6 tradesmen £7.95   = £ 47.70
Total   £ 56.15

However, the foreman is only working for 50 per cent of the time and supervising for the other 50 per cent, so the hourly rate of the gang in productive terms is:

£56.15 ÷ 6.5 = £8.64

The extra cost of supervision for the gang is:

£8.64 – £7.95 = £0.69

So the total cost per year of this supervision is:

$£0.69 \times 1869$ (the available annual hours) $= £1289.61$

Working Rule Agreement allowances

For the purposes of this exercise there are none.

Tool allowances

Under the NWRA, joiners, for example, are entitled to approximately £2.00 per week, and other trades approximately £0.50, for providing and maintaining their tools. Although not part of the basic wage, it is subject to National Insurance contributions and tax deductions.

Construction Industry Training Board (CITB) Levy

Construction industry employers are obliged to pay to the CITB a levy of 0.5 per cent on the wages and salaries of operatives, clerical workers and management if taxed PAYE, and 1.5 per cent on labour-only sub-contractors, so that training can be provided for construction skills, supervision and management. This figure is added to the labour unit costs.

National Insurance contributions

The amount of National Insurance contributions (NIC) paid by the employer and employee is determined by the Chancellor of the Exchequer in the annual budget and therefore changes from time to time. At the time of writing the employer's contribution is linked to the employee's weekly or monthly wage and is currently 12.8 per cent of gross income.

Holiday credits and death benefit scheme

The employer pays the management company the equivalent of the daily rate of each operative's wages for each day of the annual holiday earned as a result of length of service during the working year.

A tradesman paid £7.95 per hour and working 7.5 hours per day, earns a daily rate £62.63. So for the 21 days holiday the employer needs to set aside £1252. The productive period of the year is 47.8 weeks (52 − 4.2), so the cost to the employer per week is $£1252 \div 47.8 = £26.20$.

Severance pay and incidental costs

When employees either hand their notice in or are dismissed there are costs involved. In both cases it would not be unusual for there to be loss of interest in the job during the period of notice and hence loss of production. The employee could go absent leaving the employer the cost of National Insurance contributions, pensions and holiday pay. For the employee who has been dismissed there is the additional cost of severance pay. These costs will vary between companies depending upon their experience, opportunity and ability to maintain a constant labour force, but allowance for this of between 1 and 2 per cent is not usual.

Employer's liability and third-party insurance

The costs vary from company to company depending on their insurance record, the insurance company and the size and type of contract. A figure of an additional 2 per cent on the labour costs is not unusual. An alternative way is to express all insurances as a percentage of the tender price and include it as part of the overheads.

Summary

All the information above is summarised in Table 10.5 adapted from the 6th edition of the CIOB *Code of Estimating Practice* (1997). This spreadsheet can be used by the estimating department to amend the inserted figures as necessary when changes occur, such as an increase in wages and conditions agreed at national level or resulting from the Chancellor's budget.

10.8.3 Labour production rates

So far, the calculation of the labour rate is rather routine and mainly a case of inserting the current figures into a set of predetermined calculations. The real skill and expertise of the estimator is in assessing the impact of the work, site conditions and restrictions on production output and then deciding on outputs and the number of operatives needed to carry out the activity, i.e. the gang size.

The company will have acquired over the years a significant amount of productivity statistics from previous work and this is of immense value to this process. Where this is not available, a judgement has to be made from their own expertise and that of others, supplemented by technical literature. It should be remembered that technical expertise from outside the organisation is often biased in favour of their own product or system.

Table 10.5   All-in labour rate calculation

Description			Hours			
Summer period	No. of weeks	30				
	Weekly hours	44				
	Total hours		1350			
	Annual holidays (days)	14				
	Public holidays (days)	5				
	Total holidays (hours)		−171			
Winter period	No. of weeks	22				
	Weekly hours	37.5				
	Total hours		825			
	Annual holidays (days)	7				
	Public holidays (days)	3				
	Total holidays (hours)		−75			
Sickness	No. days (winter)	8	−60			
Total hours for payment			1869			
Inclement weather allowance		2%	−37			
Total hours available for work			1832			

		Tradesman	Gen. Ops	Tradesman	Gen. Ops
Earnings	Guaranteed min. wage/week	£298.00	£240.00		
	Bonus	£89.63	£72.38		
	Plus rate – attraction bonus	0	0		
	Total weekly rate	£399.63	£322.38		
	Hourly rate of pay (37.5)	£10.66	£8.60		
	Total Annual Earnings			£19,923.54	£16,067.34
Additional costs	Summer overtime hours @1.5	7.5	7.5		

continued…

Table 10.5   continued

		Tradesman	Gen. Ops	Tradesman	Gen. Ops
	Winter overtime hours @ 1.5	0	0		
	Summer overtime total hours	97.5	97.5		
	Winter overtime total hours	0	0		
	Total annual cost overtime hours			£775.13	£624.98
	Sickness pay for 5 days	£12.87/day	£12.87	£64.35	£64.35
Trade supervision*	No. tradesmen per foreman	6	0		
	Supervision plus rate/hour	£0.50	0		
	Time spent on supervision	50%	0	£1,289.61	0
WRA allowances	Tool money per hour (tradesmen)	£0.05		£93.40	
	Plus rate per hour (gen. ops)		£0.10		£186.90
	Sub total			£22,145.13	£16,943.57
Overheads	Training levy (based on LOSC)	1.5%	1.5%	£32.99	£25.23
	Employer's NI contribution	12.8%	12.8%	£2,814.98	£2152.77
	Holidays with pay	£7.95/hr	£6.41/hr	£1252.00	£1009.58
	Public holidays 8 days	7.5 hrs/day	7.5 hrs/day	£477.00	£384.60
	Sub total			£26,722.10	£20,516.18
Severance pay		2%			
	Sub total			£27,256.54	£20,926.50
Employer's liability		2%		£27,801.68	£21,345.03
Annual cost of operatives	Divide by productive hours -1832				
Cost per hour				£15.18	£11.65

*there could be an entry under Gen. Ops for a ganger

There are very many factors affecting output so the estimator must scrutinise the drawings and information provided to establish difficulties that might arise and ensure the quality specified is attained and a safe method of work is provided. Typical production issues to be looked for include:

- The quantity of work as large amounts might mean using larger, more efficient plant, and operatives become more productive as they learn all the nuances of the job.
- Repetitive work is more productive than non-repetitive work.
- Can the operatives get a good run at the work or does the sequencing of other work result in them continually being moved on and off the job.
- Is the work complex either slowing down the process or requiring tradesmen with specific skills?
- How accessible is the work in terms of access, height or depth? This can affect the ease of delivering the materials to the work place, the type of equipment that might be used and the freedom of movement of the operatives in carrying out the task.
- Heavy or irregular shaped components can cause loss of productivity.
- Restrictions on working hours, shift patterns, the need to use respirators or other restrictive protective clothing and extremes of temperature, all contribute to reductions in productivity.
- Depending upon the activity, the time of year and associated weather conditions can impact on productivity. For example, if it is very cold and windy, operatives working outside in exposed situations will inevitably spend a certain amount time trying to keep warm.

10.9 Building up material rates

10.9.1 Determining the amount of material required

Either the supplier or the contractor has to account for the fact that materials are not always ordered in the way they have been measured for the purposes of the bills of quantities and a conversion has to take place. Examples of this are bricks that are mainly measured in m^2 but are ordered from the supplier in thousands. Hardcore is measured in m^2 or m^3 but ordered in tonnes, and damp-proof membranes in m^2, but ordered in by the roll and width.

The majority of items measured using the SMM7 are measured net. This ignores items such as overlapping tiles and roofing felts that require extra material to complete the work. Materials such as concrete and hardcore lose some of their volume when vibrated or compacted. This extra material has to be calculated and taken into account when ordering. Waste also has to be accounted for and is covered in more detail in section 10.9.3. When building up the unit rate for the tender these processes are reversed and the quantities converted back to the nomenclature of the bills descriptions.

10.9.2 Checking suppliers' quotations

Whilst the estimator may have experience about the costs of materials it is important to obtain quotations for the materials to be used on the specific contract, as the transport costs will vary and prices do fluctuate for the following reasons:

- the distance or time it takes to transport materials from source to the site;
- whether or not the quantity being delivered is sufficient to take up the full capacity of the vehicle;
- large orders tend to reduce the haulage costs as the number and frequency of loads increase – this means the haulage contractor has greater certainty on achieving full use of the vehicles and can be more competitive.

Some bulk materials are relatively cheap hence the percentage cost added for haulage is significant. The exceptions to this are supply and fix by sub-contractors, and when the contractor's buyers have negotiated rates with a supplier of bulk materials such as aggregates.

When a receiving a quotation, it is checked to see that the quotation is for what was requested. Most suppliers are honourable in intent, but for a variety of reasons, including making errors or misunderstanding the nature of the request, may not quote appropriately. Certain less scrupulous suppliers want to present a low and attractive price to obtain the order and may 'confuse and confound' the inexperienced (see sections 10.15 and 10.16). Typical issues to be inspected are:

- Do the materials quoted for totally meet the specification stated in the enquiry? Sometimes the supplier proposes a lower or higher grade material than is required, whether because of supply issues or because the specifier is unaware of improvements in production and has requested a lower specification than is normally available.
- If a section of the bills has been sent for a supply-only price, has every part been priced? An example of this is precast concrete. The bill sent might include kerbs and lintels as well as architectural cladding components. These are two different manufacturing processes and it

would be unusual for the architectural cladding provider to price for kerbs and lintels and vice versa.

- Have transport costs been included? These can be high for the reasons outlined above if a supplier quotes the cost as ex-works.
- If a supplier has to deliver a part load only, it usually costs more and their conditions may state that.
- If it is expected that a delivery may arrive on site and not be unloaded immediately, because of limitations to drivers' working hours, if this becomes excessive it has many implications to the haulage company. They may well lay down maximum waiting time after which a penalty will be invoked.
- Manufacturers require time from the receipt of order before they can commence delivery, called lead-in time, and will normally state the duration. The estimator needs to check that from when the order is placed, this lead-in time will comply with the site's programme requirements.
- The supplier will usually state the duration over which the components will be delivered. It is important this complies with the site programme.
- Sometimes penalty clauses are introduced to cover for the situation when the site programme falls behind schedule. This is because the supplier may run out of space for storage or has to reduce the rate of manufacture to a less economic level.
- Some materials come in minimum sizes and are quoted at so much per unit, but the contract may require less than the amount quoted. This can either be dealt with in the price and written off as such, or taken into account in the waste calculation. If the amount not used is significant then another use elsewhere could be considered. However, it should be noted that this could incur extra costs for transport and storage not to mention any environmental considerations.
- The costs may not include all other materials required, for example nails, screws, and bolts.
- Who is responsible for unloading the vehicle? The driver may carry it out using unloading equipment attached to the lorry or the contractor may have to provide personnel to unload either by hand or aided by mechanical plant.
- If the material or component has to be tested, it is normally the contractor's responsibility and the costs incurred accordingly, but if the supplier does the testing, the estimator needs to be satisfied they have accounted for this.
- Is the price quoted a firm price or are fluctuation clauses involved?
- Discounts are often offered by suppliers normally dependent on speedy settlement of the invoice by the contractor. The relevance of this has

to do with the current inflation rates and the affect on cash flow for both parties. Higher discounts can also be negotiated if the order is for full loads delivered directly from the manufacturer rather than builders' merchants.

- Where one party is concerned about the financial state of the other, pro-forma invoices may be used. This means that the invoice has to be paid prior to goods being delivered.

10.9.3 Assessing likely waste

The SMM7 states what work is measured net and the contractor must take waste into account. The amount of waste likely to be generated during the construction process is difficult to estimate (*Operations Management for Construction*, Chapter 5). Most estimators have their own rule of thumb based on their experience, but these tend to be general figures. Different materials create different percentages of waste for several reasons, some of which are noted below along with other considerations:

- Some materials come in sizes not compatible with the dimensions required. Typical extreme examples of this are board materials, especially plaster board, pipes, timber, reinforcement steel, bricks and blocks.
- Some work cannot be practically carried out to the dimensions specified because of buildability issues. The dimension of the excavator's bucket digging a strip foundation determines the width of the excavation rather than the stated dimension of the concrete.
- There is the likelihood of damage associated with unloading, storage, double handling, fixing and other trades working in the vicinity of the product.
- The risk of loss of materials through theft and damage.
- Poor workmanship resulting in either remedial work or reconstruction.
- Using materials for the wrong purpose such as bricks in lieu of block work because it is easier for the bricklayer to do this than cut a block.
- The costs of disposal of the wasted material are rising as landfill taxes increase and segregation into different material types needs to occur before removal from the site.

With the pressures of costs and environmental implications, there is a need to consider how waste material can be reclaimed, recycled or reprocessed offering the opportunity to generate income to set aside against the cost of waste.

To improve the accuracy of these calculations, there is information available in research papers, price books and feedback from within the organisation. The problem with the latter is that contractors move on from the contract to the next and are in a 'looking forward rather than back' mode, so it is often difficult to ensure adequate feedback occurs in spite of the value of this source of information.

10.9.4 Storage, security and handling of materials

Whilst covered in the preliminaries, the site visit and site plans give the estimator an indication of problems and difficulties that might occur during the construction process. For example, a very restricted urban site may result in materials being delivered within a narrow time frame or there may such a limitation on storage that part loads have to be considered.

10.10 Cost of plant

Plant development occurs all the time at the larger end of the scale (cranes and excavating equipment) right down to hand tools, drills and saws. Some of these advances can have significant effect on productivity and costs. In small tools the development of battery-driven hand tools now have enough power and duration to be effective, has eliminated the need for a supplied power source. On larger plant, developments of hydraulic systems, laser guidance and control mechanisms have dramatically improved productivity. It is important therefore, that the estimator is well versed in these so the appropriate piece of plant can be selected for the work on the site.

There is an academic debate in many of the texts as to whether or not plant should be hired or owned, but increasingly, in a specialist area such as piling, most medium and large contractors hire plant as required so as to offer the greatest flexibility and ability to use the most up-to-date, purpose made equipment. The major exception to this is if the contract duration is of such a length as to make the economics of purchase and maintenance stack up.

On obtaining quotations for hiring plant, besides hire charges, there are several points to understand and questions that need to be answered:

*   The cost of hire may commence when the plant leaves the plant depot and finishes when it returns. These travelling costs can be significant as in the example of a large mobile crane which may take a day to get to site, only work for a couple of days and then take a day to return to its

depot. It highlights the need to consider continuity of work on site for expensive plant.

- It is usually on a weekly hire basis even if it is used for less. The exception to this is some of the large specialist equipment such as mobile cranes and daywork lorries. Continuity of work is again an issue.
- There will be contractual obligations to ensure the plant is used, maintained and worked in a safe manner. These need to be scrutinised for any onerous clauses.
- Does the plant hire include the cost of the operator and if so what are the rates and conditions of pay? There might include limitations on driver's hours.
- Even though the plant hirer supplies the operator, the contractor will have to take responsibility for any damage caused by the operator such as damage to services, unless it can be demonstrated that the operator is negligent, but even this may be excluded from the contract between the parties.
- Is the hirer or the contractor responsible for insurance? This may be split responsibility, the hirer whilst in transit to and from the site, the contractor whilst on the site.
- The hirer will take care of breakdowns and maintenance. If no operator is provided, basic maintenance such as checking oil and water levels, keeping the plant clean, inspecting the tyres, provision of fuels and oils is the contractor's responsibility.
- The plant selected has to be available for when it is required on site.
- Is the quotation a fixed price for the duration of the hire or are there fluctuation clauses? On projects certain pieces of equipment such as dumpers could be used for virtually the whole project, which could extend over two or three years.
- There may be a requirement to provide extra equipment to operate the plant. Examples are slings and lifting beams for handling components, concrete skips, specific types of excavator bucket and tools for compressors.

After this process the all-in rate for each item of mechanical plant can be calculated. In summary these are:

- The hourly cost of the machine, including the maintenance labour and parts, depreciation costs, which have to be paid for either by the hirer or the contractor owning the plant.
- Estimate the amount of likely standing or idle time. Whilst it is expected that especially heavy and expensive plant will be used to its full capacity,

in reality there will be down time resulting from mechanical failure, or inclement weather and hold ups.

- The all-in rate of the operator which is usually higher than calculated in section 10.8.2 because the working week is longer due to extra tasks such as oiling, greasing, minor repairs and coming in earlier to start the machine and position it ready for the start of the normal working day. There may also a higher basic wage rate because of enhanced skill requirements.
- The fuels, oils and greases.
- Whether or not to use this all-in rate for the plant in the preliminaries or in the unit rate in the bills.

A typical example of the all-in rate for a piece of plant shown in Table 10.6. It is assumed that a 37.5 hour, five-day basic week is being worked with overtime, the plant will be used on site for four weeks, delivery and retrieval charges are assumed to be inclusive in the hire rate. The contractor provides the driver. The driver spends 0.5 hours per day maintaining the plant that is over and above the working day. It is assumed the plant is idle on average for 1 hour per day due to stoppages caused by inclement weather and other hold ups.

Table 10.6   All-in plant rate calculation

Basic weekly hours	37.5		
Add 0.5 hours per day maintenance	2.5		
Total hours worked		40.0	
Standing time @ 1 hour per day	5.0		
Total productive hours per week	40.0 – (2.5 + 5.0)	32.5	
Weekly hire cost of plant say £200			
Productive hourly hire rate	£200 ÷ 32.5		£5.71
Fuel at 4 litres per hour @ 75p a litre			£3.00
Grease and oil @ 20p per hour			£0.20
Total weekly operator costs (40 hours)	£15.18* + 20p plant ops. plus rate	£615.20	
Hourly productive rate	£615.20 ÷ 32.5		£18.93
Total all-in cost per hour			£27.84
Total all-in cost for period on site	£27.84 x 4 x 40		£4,406.40

* see Table 10.5 tradesmen rate

The all-in cost per hour can be used for building up rates in the bills and for the period on site if the plant is to be costed in the preliminaries. This calculation does not take into account productivity issues such as the output from a back actor excavating clay being different from that of shale or the output being a function of the bucket size.

10.11 Calculating and pricing unit rates

Previously it has been demonstrated how to calculate the hourly cost of labour, how to assess the quantity of materials required, taking account of waste and how to establish the hourly rate for a piece of mechanical plant. None of these calculations takes in to account the different quantities that can determine whether or not output is high or low. In other words how long it will take to complete an operation. Other than the general comments made in section 10.8, it is not the purpose in this text to investigate in detail influences on productivity. The estimator will in practice take these factors into account when pricing the unit rate.

Estimators have at their disposal many sheets of data showing a variety outputs of various operations for differing circumstances. These cover every aspect of construction work from plant outputs in different materials to the coverage of different types of paint on various surfaces.

Adapted with permission from Brook (1998), Table 10.7 shows typical data for fixing softwood skirtings, architraves, trims, etc.

The next stage is to build up the all-in rate, which brings together the data for the labour, material and plant costs required to execute the described bills item. Table 10.8 demonstrates this for building a half-brick-thick wall using cement mortar 1:3 mix, built in stretcher bond. The table demonstrates a rate to lay 1000 bricks and also the rate to lay one metre super of brickwork.

10.12 Costing the preliminaries

Section 10.5 gives descriptions of the elements contributing to the total cost of the preliminaries and can be referred to where appropriate. Using the five categories as outlined in section 10.4, the costs are as follows.

General project details (A10–A13)

Since these are all to do with the description of the site and provide information for the tendering process, there will be no costs incurred.

Table 10.7   Output rates for softwood

Size of member		Nailed	Screwed	Plugged & screwed	Size of member		Nailed	Screwed	Plugged & screwed
19	19	0.10	0.13	0.18	25	19	0.10	0.13	0.19
	25	0.10	0.13	0.18		25	0.10	0.13	0.19
	32	0.10	0.13	0.18		32	0.12	0.15	0.21
	38	0.12	0.15	0.20		38	0.12	0.15	0.21
	44	0.12	0.15	0.20		44	0.12	0.15	0.21
	50	0.12	0.15	0.20		50	0.12	0.15	0.21
	63	0.12	0.15	0.20		63	0.15	0.19	0.25
	75	0.15	0.19	0.24		75	0.17	0.21	0.27
	100	0.17	0.21	0.26		100	0.19	0.24	0.30
	125	0.19	0.24	0.29		125	0.23	0.29	0.35

The above outputs are for each joiner and the figures are hours/metre run
ADD 30% to the outputs for fixing hardwood
Average waste allowance is 7.5%. Varies depending upon number of short lengths and mitres
ADD for cost of screws in lieu of nails

Contractual matters (A20)

Using the JCT05 Standard Form of Building Contract as the basis for the comments in this section, many of the clauses would have no cost attached them. Highlighted are the clauses most likely to result in costs to the contractor. This is not meant as a definitive list, as there is often ambiguity in that some of the clauses can be costed either here or in other preliminaries items.

Clause 17: Practical completion and defects liability

Although this can be costed elsewhere, the contractor needs to cover for rectifying the defects that occur during the defects liability period after the building has been handed over. Most contractors would take the view that these would be minor defects resulting from such causes as shrinking as the building dries out when in use or minor settlement resulting from the ground reaching a state of equilibrium after the total weight of the building has been exerted through the foundations. A nominal sum, based upon experience of the type and size of the works would be calculated based upon time for inspecting and agreeing the work to be done, and the labour and materials needed.

Table 10.8   All-in build up rate work sheet

Reference: Date: Trade: Bricklayer				Bills item description: Common brick wall, 102.5mm thick, vertical stretcher bond, 215 × 102.5 × 65mm			
Item details				*Analysis*			*Net unit rate*
Description	*quantity*	*unit*	*rate*	*lab*	*mat*	*pl*	
Materials							
Deliver to site common bricks	1000	No			150.00		
Wastage	5%				7.50		
Mortar delivered mixed	1	m³	£120				
Mortar 0.53m³ per 1000 bricks					63.60		221.10
Labour							
Bricklayers	2	hr	£15.18	30.36			
Labourer	1	hr	£11.65	11.65			
Gang cost				42.01			
Output 55 bricks/hour/brickie							
Cost = 1000 × 42.01/(2 × 55)							381.91
Total cost per 1000 bricks							603.01
59 bricks per m² for ½ brick wall							
Cost per m² £603.01 × (59/1000)							£35.58

Clause 22: Insurance of the works

Either the client or the contractor can pay this. A specialist and complex subject and normally advice would be sought from a broker. The key insurances likely to be taken out would be:

- Employer's liability Insurance, which is mandatory, covers claims arising as a result of death or injury of an employee. This is calculated based upon the company's payroll and so can be added into the all-in rate for management and labour.
- Public liability insurance whereby the contractor insures against any claim arising from death, injury or damage to third parties caused by the

negligence of the contractor's employees. This is also calculated based upon the payroll and can also be added into the all-in rate.
- Joint liability insurance, taken out by both the client and the contractor to cover eventualities such as damage caused to third parties' property during the course of the construction works.
- The contractor may be asked to take out insurance to cover the building works and materials against all risks. This is based upon the total contract value, any assumptions made for increases of costs from the original tender and administration costs and the fees of the insurance broker.

Clause 30: Certificates and payment

These are the conditions regarding the percentage of retention the client is going to deduct from the monthly valuation of work done and when this will be returned, so this is not a direct cost to be entered here. It has an indirect impact on the amount of monies to be raised by the contractor to maintain sufficient funds to pay sub-contractors and suppliers until the contract becomes profitable. This will affect the amount of overheads added to the tender price.

Clause 38: Contribution, levy and tax fluctuations

Sometimes referred to as the fixed price option. This is only to take account of increases of costs to the contractor of such items as National Insurance contributions, taxes, landfill taxes and levies on which they have no control, but does not include for inflationary increases to labour materials and plant. These would be covered elsewhere based upon a method developed by the National Economic Development Office, but are seldom applied. Here the contractor is asked to predict the likely increases in building costs.

Employer's requirements (A30–A37)

A30 Tendering/subletting/supply

The financial consideration here is if the employer requires a guarantee bond to be taken, usually up to 10 per cent of the value of the contract (section 9.5.3).

A31 Provision and content and use of documents and
A32 Management of works

There is no cost to be entered here. The latter would be taken into account in the general overheads.

A33 Quality standards/control

The provision of constructed samples and components can be costed in the usual way and entered here. If samples of concrete are to be carried out this can be readily calculated. For example, the numbers of core samples and tests, cube tests, etc. can be ascertained from the instructions in the employer's requirements in conjunction with the amount of concrete ascertained from the bills, gives the data required. The local concrete-testing laboratory will have a scale of fees for each test.

A34 Security/safety and protection

Safety would not normally be priced here as it would be included elsewhere either as an overhead component, a staff component in A40 or as part of the bills item. Security, such as the provision of hoarding, patrolling and alarms is normally found here. A simple example is shown in Table 10.9.

A sum may also be added to cover maintenance of external roads and services, noise and pollution control, protection of trees on the site, protection of works in progress and repairs to surrounding buildings. How much, if any, is a function of the risk the contractor assesses there to be.

Table 10.9 Security costs

Description	Cost (£s)
Hoardings around the site	5,000
Repairs to the same	750
Chain-link fencing around secure compound	600
Gate house and gates	2,000
Alarm system	700
Night security patrol for contract duration (15 months)	15,000
Total	£24,050

A35 Specific limitations on method/sequence/timing/use of site

These limitations are considered and allowances made for in the methodology of construction and the build-up of the all-in rates. Where this is not possible a sum might be added.

A36 Facilities/temporary works/services

This is usually taken account of in A41, Site accommodation.

A37 Operations/maintenance of the finished building

If this were solely the provision of plant manuals and operating instructions, there would be no need to cost it. However, if this involves the training of operatives in the use of plant, allowances are made here. Factors affecting the costs include the staff, the duration of the training, travel and subsistence.

Contractor's general cost items (A40–A44)

This is where the serious money is spent in the preliminaries, so it is important that fixed and time-related charges are identified.

A40 Management and staff

Taking account of overall costs of employing staff as outlined in section 9.5.4, the estimator needs to establish the cost of each member of staff working on the contract. This depends on the duration of their employment on site. A standard format can be used to list all possible members of staff, but for the purposes of this example, as shown in Table 10.10 only those employed are shown. Under column 'Number', the amount of time they will be spending on *average* each week over the duration of the contract is assessed. In some cases, such as the engineering staff, there may be weeks, especially at the beginning of the contract and when the structure is being set out and constructed, they are employed full time on site, but then their presence is not required until the setting out of the site works.

Whilst this gives the overall predicted costs of management and staff, which is fine for tendering purposes, it does not demonstrate when staff are required on the contract. Since staff costs are considered to be time related, these data need to be calculated to demonstrate the monthly costs. For the purposes of this example it is assumed, for simplicity, all the staff are paid on

Table 10.10   Site staff costs

Staff description	Number	Salary* (£/week)	Total cost (£/week)	Expenses subsistence, etc.	Cars £150/ week	Total (£/week)
Contract manager	1	700	700	100	150	950
General foreman	2	500	1000	2 × 200 = 400		1,400
Trades foreman	3	450	1350	3 × 200 = 600		1,950
Ganger	1	400	400	200		600
Engineer	0.2	650	130	0.2 × 100 = 20	30	180
Assist. engineer	0.2	400	80	0.2 × 100 = 20	30	130
Chainman	0.2	350	70	0.2 × 100 = 20		90
Planner	0.5	650	325	0.5 × 100 = 50	75	350
Quantity surveyor	0.5	650	325	0.5 × 100 = 50	75	350
Safety officer	0.1	600	60	0.1 × 100 = 10	15	85
Secretary	1	300	300			300
Typist	0.6	250	150			150
Cleaner	1	200	200			200
Total cost/week						£6,735
Total cost/contract				10 months or	44 wks	£296,340

* salary includes for employer's contribution to pensions and national insurance, CITB levy, etc.

a monthly basis. Under each monthly column in Table 10.11 is the total cost of each person to the contract, based on 4.33 weeks per month.

A41 Site accommodation

The assumption is that all the accommodation required for both the contractor's and client's requirements are costed here. After the initial set up costs and dismantling costs, the accommodation provided is rented on a monthly basis over the period of the contract. This identifies the time related and fixed charges. Table 10.12 represents a typical calculation excluding furniture.

A42 Service and facilities

Table 10.13 demonstrates the breakdown of the various services typically considered when pricing this section. The costs per item have not been shown since they are extremely variable depending upon the particular site

Table 10.11 Monthly site staff costs (£)

Staff description	Jun	Jul	Aug	Sep	Oct	Nov	Dec	Jan	Feb	Mar
Contract manager	4113	4113	4113	4113	4113	4113	4113	4113	4113	4113
General foreman	6062	6062	6062	6062	6062	6062	6062	6062	6062	6062
Trades foreman	2814	2814	5629	11258	11258	11258	11258	11258	8442	8442
Ganger	2598	2598	2598	2598	2598	2598	2598	2598	2598	2598
Engineer	1559	1559	1559	779	779				779	779
Assist. engineer	1126	1126	1126	563	563				563	563
Chainman	779	779	779	390	390				390	390
Planner	1516	2273	2273	2273	1516	1516	1516	758	758	758
Quantity surveyor	758	758	1516	1516	1516	1516	1516	1516	2273	2273
Safety officer	368	268	368	368	368	368	368	368	368	368
Secretary	1299	1299	1299	1299	1299	1299	1299	1299	1299	1299
Typist				1082	1082	1082	1082	1082	1082	
Cleaner	866	866	866	866	866	866	866	866	866	866

Table 10.12 Site accommodation

Fixed charges	Cost (£)	Time-related charges		
Description		Description	Rental/ month (£)	Total cost (£)
Prepare foundations	500	Staff offices	1,000	10,000
Transport and delivery	400	Canteen/mess room	500	5,000
Crane hire	150	Materials store	300	3,000
Installation	200	Clerk of works' office	200	2,000
Service connection	65	Washing facilities	300	3.000
Dismantling	200	Toilets	250	2,500
Total installation	1,515	Security office	200	2,000
Crane hire	150	Total monthly cost	2,750	
Transport and return	400	Total contract cost		£27,500
Dispose of foundations	250			
Total removal	800	Total accommodation costs		£29,815

and authority. There are fixed and time-related charges. Some items such as electrical wiring and cleaning comprise both labour and materials.

The provision of services for the client's representatives can either be included here or separated out depending upon the contractual requirements. Some contractors include attendance allowances in this section, but with the increased use of sub-contractors there is a growing tendency to add these to general site overheads, as many of the facilities provided are used by all.

A43 Mechanical plant

This is for plant not included in the bills rates. There are two distinct categories to consider. First, tower cranes and hoists, etc., which have both a fixed charge for installation and dismantling and time-related when in use, and those such as dumpers, which are purely time related.

A44 Temporary works

The provision of access roads and scaffolding are the main items included here. Items such as strengthening roads, temporary walkways and traffic control are not necessary on every site and are calculated as required. Signage can be priced here and would be a lump sum, but can be included in general overheads instead.

Table 10.13   Services and facilities

Fixed charges		Cost (£)	Time-related charges	Cost (£)/ mth
Electricity	Connection charge		Electricity charges	
	Wiring		Site temporary electricity supply	
Telephone	Connection charge		Calls and line rental	
	Answering machine			
Water	Connection		Water charges	
Rates	Temporary accommodation			
Fire	Extinguishers		Extinguisher maintenance	
Office	Computers		Stationery and postage	
	Fax machine			
	Photocopiers			
Waste			Chutes	
			Skips	
Cleaning	Office if not included in A40		Roads	
			General site cleaning	
			Cleaning building on handover	
Protection			Protection of work in progress	
Small tools			Approx. 0.2% of contract value. Can be in A43 instead	
Safety	Signs		Protective equipment	
			Training	
Security	Alarm installation		Staff if not included in A40	

The cost of access roads varies considerably from site to site depending on its length, amount of traffic usage and the quality of the sub-soil upon which it is laid. A detailed look at the materials available and design criteria are explored in *Operations Management for Construction*, Chapter 1. It would be considered a fixed charge, as it is a single event, hopefully requiring little maintenance, if designed properly. A monthly allowance can be included to take account of minor repairs and cleaning involving both labour and materials. The costs will include for preparing the sub-soil, providing and laying the material and occasionally the provision of drainage.

The majority of scaffolding is carried out by a sub-contractor who gives a price for both erection and dismantling, and then charges a hire rate for the intervening period. Minor modifications, such as the removal and replacement of handrails to allow the passage of goods into and out of a building or the removal of ladders at the end of the working day, may be carried out by the main contractor, providing liability issues are resolved.

If the contractor is carrying out the scaffolding, the material needed is hired from the supplier and then erected by the contractor. The weekly or monthly hire charge is based upon the length of scaffolding tubes, numbers of fittings and length of boards ordered. It will also include an extra for delivery and collection.

Work carried out by others (A50–A55)

A50 Work/materials by the employer

The cost to the contactor is the attendance allowance required to assist the client in carrying out the work or unloading their materials.

A51 Nominated sub-contractors

The estimator adds profit and an attendance allowance to each of the nominated sub-contractor's prime cost (PC). The profit, expressed as a percentage, does not have to be the same as added to the builder's work. When selecting the profit margin it should be noted there is probably less risk associated with those nominated as the onus is on the client in terms of the reality of their price. This can be taken into account when deciding on the amount to be added.

The attendance allowance may vary between sub-contractors depending upon the amount required. It may be specified that the contractor has to provide special attendance such as the provision of a mobile crane and its operative for fixing precast cladding panels or the steel structural frame. This can be easily calculated based on hire rates and predicted duration. Other than any special attendance, attendance for provision and use of general facilities is added as a percentage of the PC sum.

A52 Nominated suppliers and A53 work by statutory authorities

A percentage profit is added to both of these.

A55 Dayworks

The procedure for calculating daywork is to record the amount of labour, materials and plant that has been used to carry out an operation. This record is signed by the architect or clerk of works to confirm the details recorded are correct. At a later stage the client's QS and the contractor can agree some other method of valuation and the daywork sheet discarded. If the daywork procedure is continued, the data is priced. This has to comply with the definitions prepared for building works as published jointly by the by the Royal Institution of Chartered Surveyors (RICS) and Construction Confederation.

The definition for building works of prime cost of daywork states that the component parts that make up a daywork are: labour, materials and plant. The contractor then adds for any incidental costs, overheads and profit at the tender stage. The effect of doing it then is it creates competition for daywork.

The prime cost components are calculated as follows.

Labour

Labour is expressed as an hourly rate, but differs from the all-in rate labour rate as worked out in section 10.8.2 in so far as many of the cost components are omitted, but taken into account in a percentage addition made by the estimator later. The hourly prime costs are calculated by establishing the annual labour prime cost and dividing by the number of working hours per annum. The annual prime cost comprises:

- the guaranteed minimum week earnings
- differential payments for skill and extra responsibility
- public holidays
- Employer's National Insurance contributions
- the CITB levy
- annual holidays and contributions to death benefits.

Table 10.14 demonstrates the calculation for general operatives and tradesmen. Note that the all-in rate calculated in Table 10.6 is £15.09 for tradesmen and £11.58 for general operatives compared with £10.50 and £8.47 respectively for the prime costs.

The percentage addition incorporates the following incidental costs, some or all of which will be added depending on the nature of the daywork activity:

- head office charges
- site staff and site supervision
- additional costs of overtime
- bonuses and incentive payments
- subsistence allowance, fares and travelling
- time lost due to inclement weather
- sick pay
- third-party and employer's liability insurance
- redundancy payments tool allowances
- use of scaffolding
- provision of protective clothing, lighting, safety, health and welfare facilities and storage
- variations to the basic rate of pay
- profit.

The calculation of these depends on the size and nature of the site and is beyond this text. Interested readers are referred to Smith (1999). In

Table 10.14 Labour prime cost rate

Prime cost of labour		Hours	
Working hours per week		37.5	
Working hours per year (37.5 × 52)			1950.0
Annual holidays (days)		21	−157.5
Public holidays (days)		8	−60.0
Total working hours per annum			1732.5

Wage calculation	Trades	Gen. ops	Trade	Gen. ops
Guaranteed weekly wage	298	240		
Annual costs for working hours	46.2 weeks		14,322.00	11,550.00
Annual costs public holiday	1.6 weeks		496.00	400.00
Extra for skill	None			
Sub-total			14,816.00	11,950.00
NI contributions	12.8%	12.8%	1,896.45	1,529.60
CITB levy (based on LOSC)	1.5%	1.5%	222.24	179.25
Annual holidays & death benefits	7.95/hr	6.41/hr	1,252.12	1,009.58
Annual prime cost of labour			18,186.81	14,668.43
Hourly prime cost rate (÷1732.5)			£10.50	£8.47

practice, the percentage addition can exceed 100 per cent of the prime cost rate.

Materials

The prime cost of materials is the invoice cost after deducting trade discounts in excess of 5 per cent and the cost of delivery if ex-works. The cost of unloading, storage and handling will be included in the labour calculation. The percentage addition is based upon those items listed above under labour percentage additions, which are considered applicable. The percentage added will be normally in the range of 15 to 25 per cent, as it comprises only overheads and profit.

Plant

The prime cost of plant excludes non-mechanical tools and scaffolding, the latter being included in the labour percentage addition. The cost of the plant operative is costed as a labour prime cost. There are two methods of assessing the value of plant.

- The RICS schedule of basic plant charges. This is a regularly updated published list of the hourly hire rates for a variety of mechanical and non-mechanical plant. Extremely comprehensive, it includes the majority of plant likely to be encountered on a construction site. The schedules also include for fuel, oils, grease, maintenance, licences and insurance.
- Current rates of hire. If the contractor has to hire plant, currently not available on site, he is entitled to be reimbursed at the rates charged by the hiring company irrespective of the rates stated in the RICS schedule.

The percentage addition is calculated based upon the applicable incidental costs listed under labour prime cost. Since in the daywork calculation, the plant is only costed for the time it is in operation, the estimator needs to take account of the time lost due to inclement weather and standing time as well as overheads and profit. Depending upon the piece of plant this percentage addition could be anything from 20 to 50 per cent.

10.13 Sub-contractors' and suppliers' quotations

Since in today's marketplace the majority of work is sub-contracted out either in the traditional way or as packages, it is essential the selection of both sub-contractors and the supplier of materials is carried out expertly.

There are four types of sub-contractors: nominated, named, domestic and labour-only sub-contractors.

10.13.1 Nominated sub-contractors and suppliers

Nominated sub-contractors are those who have been selected by the architect under Clause 35 of the JCT05 to supply and fix materials as part of the contract. The architect will have usually gone out to tender and obtained a quotation for the work at an earlier stage. One of the main reasons is the lead-in time for commencement of this work from receipt of order is such that if left to when the contract is awarded, it would cause delays to the construction process. Traditionally specialist sub-contractors such as lift installers and mechanical and electrical service engineers were nominated, but the practice is less common than it was. Clients may also insist upon a nomination if, for example, the work could be carried out by one of their subsidiary companies.

The contractor can make 'reasonable objection' to the nomination in writing usually based on previous experience of working with the sub-contractor such as their quality of work, inability to keep to the programme or unwillingness to work in a team. If these can be demonstrated to be valid, the architect would be silly to ignore these comments.

There are risks in nominating because if the sub-contractor delays the project as a result of their performance on site, or if there are defects or delays in their design, the contractor can legitimately claim for an extension to the contract and would not have to pay any contractual penalties for late handover as a result. Further the contractor has no incentive to ensure the sub-contractor completes the work on time and could manipulate the situation to use the delay to cover up his own inefficiencies elsewhere on the project. In today's current climate of co-operation, many contractors find the lack of control of both the design and productivity sometimes to be a hindrance in completing the contract on time. On the other hand the contractor not being responsible for the work can still, generally expect to receive overheads, attendance and profit applied to the prime cost sum. It is not always the case, but the risk relative to employing sub-contractors direct is lessened.

Examples of why a supplier is nominated is the selection of an architecturally influenced material such as ceramic tiles, which can only come from one source or a material where the lead-in time may be problematic. Finally there are situations where the nominated supplier or sub-contractor offers a near unique service especially if this involves a design component.

10.13.2 Named sub-contractors

This is used especially if the form of contract selected is the JCT05 Intermediate Building Contract (IFC05) as there is no provision in this contract for nominated sub-contractors. Still chosen by the architect, unlike with nominated sub-contractors, the contractor is required to take responsibility for the sub-contractor's performance both in terms of delays and quality. The main contractor is also responsible for certifying payments.

The architect invites tenders from chosen sub-contractors and after selection, if this occurs prior to the preparation of the main contractor's bid, the named sub-contractor's tender is given to the contractor who is required to include it in the main contract and not as a PC sum. This means the contractors can add any percentage they like to cover attendance, overheads and profit. Unlike with domestic sub-contractors, the main contractor is not in a position to renegotiate the price with named or nominated sub-contractors, if their bid is successful.

Another method is the architect gives a list of named sub-contractors, from which the main contractor can choose one to include in their bid. They may be allowed to suggest alternatives to the architect to be included on this list. In this case, in spite of the 'named' list, the contractor makes the contract with the sub-contractor so these are classified as domestic sub-contractors.

10.13.3 'Domestic' sub-contractors

Whether using the JCT05, IFC05 or most other forms of contract, the majority of sub-contracted work falls into this category. The main contractor is responsible for the final selection and engagement of the sub-contractor and has full responsibility to ensure the work sub-contracted is carried out properly and to programme. Theoretically in the conditions of most contracts, the contractor can only sub-let work with the architect's written consent, but the clauses go on to say that this should not unreasonably be withheld. Such reasons are more likely to be based upon the client's ethical beliefs and requirements.

The main reasons for the main contractor to sub-contract are as follows:

- The reallocation of risk by shifting all or part of the risk to another party.
- To overcome the problem of finding continuous employment for full-time employees when the work load fluctuates as this may be difficult or uneconomical, which in turn can give more flexibility in planning the works.

- To supplement the contractor's own labour force to overcome peak loads thereby maintaining or accelerating the programme.
- To deal with specialist work, such as piling, structural steel, mechanical and electrical services.

It does mean the management changes from managing one's own staff to that of co-ordinating packages. This is a different emphasis. It means there is potentially a loss of control especially if the sub-contractor sub-lets part or all of the work. The main contractor may wish to prevent this by having a clause prohibiting this. There is also an argument that the amount of profit that can be made by the main contractor is limited as they can only make the percentage added to the sub-contractor's quotation and cannot benefit from any productivity gains made by the sub-contractor. The only other opportunity to earn more is renegotiating the sub-contractor's tender on being awarded the main contract.

As with the client, contractors have their own preferred lists of sub-contractors and suppliers upon whom they can rely upon for all the reasons as follows:

- their reputation for good quality workmanship
- the quality of their management
- completion on time
- their experience on similar work
- their financial state
- the experience the contractor has had with the sub-contractor on previous work.

The domestic sub-contractors as outlined above are those that supply and fix all the necessary labour, materials and plant. This will usually involve providing some site accommodation such as toilets and washroom and, on large sites, canteen facilities.

10.13.4 Labour-only and labour- and plant-only sub-contractors

Labour-only and labour- and plant-only sub-contractors provide, as the titles indicate, only labour, or labour and plant, the materials being provided by the main contractor. Typical examples of the former are brickwork and concreting gangs, and the latter, a reinforcement steel sub-contractor who may provide the cutting and bending equipment.

They are selected and engaged in the same way as domestic sub-contractors, and the main contractor has the responsibility to ensure quality and programme commitments are met. The contractor needs also to be satisfied that the labour-only sub-contractor has adequate insurance and that they hold a Construction Industry Scheme (CIS) certificate to satisfy the HM Revenue and Customs (HMRC) and are not working on the black market.

The reasons for employing labour-only are similar to that outlined in domestic sub-contractors (section 10.13.3), but there are other issues to be clarified before signing any contract. These include:

- Are any attendances required especially the use of plant if not provided as part of the contract?
- Are they or the main contractor responsible for unloading their materials and does this include moving from storage to the point of production?
- Who is responsible for setting out their work?
- Is the sub-contractor providing its own supervision to ensure quality, safe working practices and keeping the work on programme?
- Are they to be paid weekly or after the monthly valuation of work is done?
- Does the contractor or sub-contractor carry the burden for retentions? If the latter than at what point is it released? It would be unfair for a small contractor to have to wait 12 months after the main contract has finished before receiving any retention monies.
- Does the contractor collect the CITB levy from the sub-contractor or do they pay it directly themselves? If the latter the contractor needs to be satisfied that this is happening.

Construction Industry Scheme Certificate (CIS)

The Inland Revenue (now HMRC) became concerned that much of the labour employed in the construction industry was labour-only sub-contracted and there was a significant gap in the legislation that permitted operatives to work, be paid and then not pay tax. The initial solution was the 714 scheme, but this was replaced in August 1999 by the CIS.

Construction companies must only employ sub-contractors who are registered under the scheme. Before the sub-contractors can be paid under the scheme they must hold either a registration card or hold a Sub-contractors Tax Certificate. To obtain either of these the sub-contractor has to register with HMRC. Sub-contractors who meet certain qualifying conditions are issued with Sub-contractors Tax Certificates. Those who do not, usually individuals or small gangs, will be issued with registration cards.

In the case where a sub-contractor holds a registration card, the contractor must make a deduction from all payments for labour of an amount to cover the sub-contractor's tax and National Insurance contribution (NIC) liability. Where the sub-contractor holds a Sub-contractors Tax Certificate, the contractor will pay the gross earned, without deductions. Full details of the CIS scheme can be found on the HMRC website: www.hmrc.gov.uk/cis/cis-intro.htm.

10.14 Selection and invitation to tender – sub-contractors

The procedure for selecting a sub-contractor is very similar to the client selecting a contractor. It is increasing likely that the sub-contractors selected to tender will either be on a preferred list, which means much of the checking procedures will have already been done, or have an supply chain agreement with the contractor (*Operations Management for Construction*, Chapter 7). If neither of these exist or if the contractor wishes to extend the preferred list, a series of questions need to be answered before giving the sub-contractor the opportunity to tender, as listed in section 10.13.3.

It is essential to obtain as accurate a price as possible from the sub-contractor. This can only happen if the information submitted to them is as comprehensive as possible. The information they require is similar to what the main contractor needs for its tender plus issues specific to the sub-contractor such as:

- basic details about the project such as location, names of employer and consultants;
- a general description of their work;
- other appropriate pages from the preliminaries including information about access and any other restrictions as stated such as working hours, noise levels, etc;
- where further information like drawings and other relevant reports can be inspected;
- the relevant extracts from the bills of the measured work to be priced by the sub-contractor with the relevant specifications and schedules;
- drawings needed to permit a tender to be produced;
- the type of contract they will enter into if successful and any amendments identified – this may well be a contract document developed by the contractor for their own use with sub-contractors;
- what attendance facilities the contractor is to provide – for example, temporary accommodation, scaffolding, unloading and storage of equipment, day-to-day setting out and provision of waste skips;

- the programme requirements and any apposite method statement;
- the date and time by which the quotation has to be returned;
- financial matters, including the method of measurement and frequency of payments, the amount of retention and defects liability period, fluctuations if any, and the daywork schedule of rates;
- either a request for discounts offered or a requirement of a percentage discount payable to the contractor;
- the contractor's health and safety plan.

10.15 Opening of tenders

Whether or not these tenders should be as sealed bids depends upon circumstances. If it is for a local authority direct works department (they are in effect contractors) then this would be required, and all bids would be opened at a given time under controlled supervision to demonstrate that no preferential treatment or fraud occurs. However, in a general contractor's office, whilst it is in the contractor's own interest to ensure the process is open and fair, the routine is usually less rigorous.

On receipt of the quotations there are two prime processes to carry out: to ensure the comparison is on a like-for-like basis; and to select the most appropriate quotation. This is not just comparing each competing sub-contractor's 'total' price. It is normal procedure to enter information of each submission on a form such as demonstrated in Table 10.15, so a true comparison can be more easily made, as it is important that all the submissions are inspected carefully. However, before analysis, the figures are checked to see that the sub-contractor has quoted as required. Issues to look for are:

- Have all the items been quoted for? The sub-contractor may consider that part of the bill section sent does not come within their remit and expertise. For example, the masonry section of the bill may be have been sent in total and includes some dry-stone walling. One sub-contractor specialising in brick and block work does not price this part, whereas another may have access to specialist dry-stone wallers and has priced the section of the bills in its entirety.
- Have they made any specification changes? This is a common ploy as by offering an alternative it can create a competitive edge over others. The revised specification may be a satisfactory solution, but at this point in time it is essential to compare like with like. It does give the contractor the opportunity to renegotiate later as well as assist in value engineering the project.

Table 10.15 Comparison of sub-contractors

Sub-contractor comparison

Project: Jasper Winston offices

Sub-contractor type: carpet tiles

Bill item	Quantity	Firm A		Firm B		Firm C	
		Rate (£)	Total (£)	Rate (£)	Total (£)	Rate (£)	Total (£)
M50/1/A	450m²	24.50	11,025.00	24.25	10,912.50	23.75	10,687.50
M50/1/B	1000m²	23.75	23,750.00	23.90	23,900.00	25.00	25,000.00
M50/1/C	630m²	34.60	21,798.00	35.00	22,050.00	35.20	22,176.00
M50/1/D	350m²	29.25	10,237.50	27.60	9,660.00	29.00	10,150.00
Total			66,810.50		66,522.50		68,013.50
Discount		2.5%	1,670.26	1.5%	997.84	2.0%	1,360.27
			65,140.24		65,524.66		66,653.23
Lowest net			65,140.24				

- Have they added or subtracted any contract conditions? It is not unusual for sub-contractors to have their own schedule of conditions and include these with their tender in spite of being asked to quote based on the contractor's conditions.
- Whilst the contractor has specified the attendance allowances to be provided, in practice sub-contractor's needs may not be fully understood so the tender may include extra provision.
- The letter of enquiry will have stipulated start and finish times for the sub-contractor's work. It is necessary to check they have accepted these and not modified them in any way.

Having investigated the above and resolved any matters, the data can then be entered. As can be seen by this example in Table 10.15, firm B appears at first to be the most competitive, but on analysis after the discount rates are taken into account, firm A is the lowest. The breakdown of the various bills items could be useful once the contract has been awarded to act as a basis for negotiation. Why for example, does firm B have a lower rate for items M50/1/A and D than the overall lowest net price of firm A? Alternatively, firm A was more expensive than firm B, before the discount was deducted. Is this negotiable?

10.16 Suppliers' quotations

On requesting quotations from a material supplier the list of requirements is stated to enable them to put together an informed quotation, although this is not as large as for the sub-contractor. The information needed includes:

- the name and location of the contract to which they have to deliver;
- any restrictions on delivery, such as time and parking;
- a description of the materials, and most important, the specification;
- the approximate total quantity needed for the contract;
- drawings of components if they are to manufacture;
- the schedule of components, for example doors and windows;
- the approximate frequency, timing and quantity of delivery – this can only be indicative as the contract start date will not necessarily be fixed yet, nor are all the designs complete;
- date by which to return the quotation;
- whether or not it is a fixed or fluctuating price;
- to state any discount terms being offered;
- it would also be useful to know if they could offer a different material, which meets the specification requirements, but is cheaper.

On receipt of the quotations they are checked in line with the summary shown in section 10.9. Suppliers often have a standard set of conditions printed on the rear side of their quotation stationery and this needs to be checked for any unacceptable conditions. Once this has been done the various quotations can be entered on a form for comparison and analysis, as shown in Table 10.16. For simplicity this table assumes all conditions of sale are the same and no discounts are offered from any of the suppliers.

For the purposes of pricing the bills it is assumed the rates can be used in isolation, picking the cheapest from each supplier, unless the supplier has specifically stated that each price is subject to all materials being ordered. If awarded the contract, and the prices remain the same, the contractor is in a position to negotiate a lower price, not just because the order is now in the bag, but because it may be in the interest of the suppliers and contractors to obtain similar materials from the same source. For example, the total price for firms A and B are similar, so would firm A be prepared to lower its price for hardcore to match that of firm B?

10.17 Method statements

Method statements are a written description of how specific activities or groups of activities are to be carried out. Increasingly, clients require method statements from contractors to ensure their knowledge of production, safety and quality. Contractors can be reluctant in open tender to submit their detailed method statements at this stage as if not successful the client could use their ideas with the successful bidder. As a result, the method statement offered to the client may be less detailed and more of a public relations exercise, designed to demonstrate the contractor's competence and experience. On the other hand, for the experienced client it demonstrates the contractor's ability to comply with its needs.

There is a more fundamental reason for method statements: to establish the best method for the needs of the contract and also as part of value engineering. The method adopted may be determined by the duration given to complete the contract and is therefore linked to the construction programme time. It might be also a useful tool to demonstrate to the client that for a small sum more, the contract could be completed much faster, giving the client the opportunity of increased revenue. Finally, it provides the basic information to enable the pre-tender programme to be produced.

Analysing the work carefully at this stage can resolve many later issues, as it could be that the method adopted alters the proposed sequence of construction requiring design information to be produced in a different

Table 10.16 Materials comparison sheet

Materials comparison sheet

Project: Jasper Winston offices

Material	Specification	Quantity	Unit	Firm A Rate	Total	Firm B Rate	Total	Firm C Rate	Total	Rate selected
Hardcore		2,358	tonne	18.3	43,151	18.1	42,680	18.8	44,330	18.1
Sand		566	tonne	14.8	8,377	14.9	8,433	14.9	8,433	14.8
			Total		51,528		51,113		52,763	

Material	Specification	Quantity	Unit	Firm D Rate	Total	Firm E Rate	Total	Firm F Rate	Total	Rate selected
Concrete	C.75: 20mm	1,324	m³	69.1	91,488	70.1	92,812	68.9	91,224	68.9
	C25: 10mm	677	m³	73.3	49,624	72.3	48,947	72.4	49,015	72.3
			Total		141,112		141,759		140,239	

Material	Specification	Quantity	Unit	Firm G Rate	Total	Firm H Rate	Total	Firm I Rate	Total	Rate selected
Masonry	Facing brick	90,000	1,000	249	22,410	249	22,410	249	22,410	249
	Eng. brick	6,000	1,000	194	1,164	174	1,044	180	1,080	174
			Total		23,574		23,415		23,490	

Table 10.17 Method statement

Item	Method	Quantity	Labour	Material	Plant	Rate	Duration	Comments
Reduce piles to datum level	Compressed air breakers	100m³	2 labourers	–	Compressor and 2 breakers	4 manhours per m³	12.5 days	Noise levels to be controlled

sequence and it can be a great starting point for the management team when planning the construction of the contract.

The first stage is to analyse the operation in question. To do this the estimator and advisors need to look at the overall site conditions such as limitations of space, access, the footprint of the building, availability of labour, requirements for phased handover of the works and so on. They will consider the climate implications such as high winds that may mean tower cranes are not used because of the amount of down time predicted. They will make note how other operations will affect each other. Finally its size and scale are taken into account, as for example, large-scale operations would allow higher capacity equipment to be used than employed on smaller operations.

Finally, a summary of the decision can be produced and Table 10.17 is a typical way of presenting the information giving the planner the resource information needed to produce a plan. An alternative form of presentation, often used for the client's benefit, is to produce the same information in a narrative format rather than tabular.

10.18 Tender programme

The tender programme is required to establish whether or not the project can realistically be completed within the time prescribed by the client and to establish the method and sequence of operations from which other issues flow and calculate the amount of resources needed to complete the work. Up to a certain limit the amount of effort put into the production of plan and method statement relates to the accuracy and competitiveness of the tender price. Its production demonstrates the contractor's intention at the time of tender and the basis upon which the tender was priced. Some clients require a programme to be produced as part of the tender submission and this document can be used also for this purpose. How to produce a programme can be seen in *Operations Management for Construction*, Chapter 2.

The programme enables the estimator:

- to establish when sub-contractors and suppliers are needed so this can be incorporated in the invitation to tender documents;
- act as a basis to establish when staff are required for the contract so site overheads can be calculated (preliminaries);
- establish when plant is required on site, giving the planner the opportunity to better resource its use;
- identify building works likely to be effected by weather conditions, such as ground works in winter and high wind;

- establish when temporary works, such as scaffolding and falsework, are required for the preliminaries pricing;
- identify possible savings in time that may be offered to the client;
- it can also show an allowance for public and peak annual holidays;
- it is an ideal mechanism for calculating the cash-flow predictions.

10.19 Cash flow

One of the important pieces of information the senior management team requires when deciding how much overheads and profit to add to the tender is how the cash flow is predicted to contribute to the business as a whole. It is normal to expect a negative cash flow at the beginning of the contract and then become positive towards the end of the project. They look at the overall cash flow of the business and determine if there are any periods in the financial year when the business is in or approaching deficit so that, if required, a financial loan can be sought.

To produce the cash-flow predictions, the tender programme and bill costs are brought together to produce a financial graph showing how much income the contract generates each month and when the contractor will receive monies from the client. Added to this are the outgoings to the suppliers and sub-contractors. Since the bills are produced in the way that they are (SMM7) it is relatively simple to abstract the costs of each stage of the programme, as each bar line on the tender programme is similar to the structure of the SMM7.

All that has to be done is to divide this sum approximately equally along each bar line in weekly or monthly increments, and total up each period's expected contribution, not forgetting the preliminaries that are fixed or time related, and deduct the retention percentage, in this case 10 per cent, as shown in Table 10.18.

The entries for the payment out and the cash flow are not completed in the table. There are other considerations to take into account such as the date monies are received, invoices paid and retentions withheld. A more detailed discussion of this subject is given in Chapter 13.

10.20 Overheads and profit

Whilst the preliminaries cover the site overheads, there is still the matter of the regional and/or head office overheads. These overheads comprise:

Table 10.18 Monthly income

Month	1	2	3	4	5	6	7	8	9	10	11	12	13	14
Activity														
A	2.8	2.8												
B		4.3	4.3	4.3	4.3									
C		7.7	7.7	7.7	7.7	7.7	7.7							
D				5.5	5.5	5.5	5.5	5.5						
E		8.4	8.4	8.4	8.4	8.4	8.4	8.4						
F						1.8	1.8	1.8	1.8	1.8				
G						9.8	9.8	9.8	9.8	9.8	9.8			
H								3.3	3.3	3.3	3.3	3.3	3.3	3.3
J								6.7	6.7	6.7	6.7	6.7	6.7	6.7
K									5.1	5.1	5.1	5.1	5.1	5.1
L							7.5	7.5	7.5	7.5	7.5	7.5	7.5	7.5
M									12	12	12	12	12	12
etc.														
Preliminaries – time related	2.2	3.8	5.6	6.1	6.1	6.8	7.3	7.0	7.8	7.8	7.6	5.4	5.4	5.4
Preliminaries – fixed	89													
Gross monthly income forecast	94	27	26	32	32	40	48	50	54	54	52	etc.		
Less retention 10%														
Net interim value	85	24	23	29	29	36	43	45	49	49	47	etc.		
Payments sub-contractors & suppliers														
Balance/cashflow														

- The salaries and costs of the directors and staff employed at the head/ regional office – this figure includes all the additional employment costs, such as employer's pension and NIC, holidays, and the CITB levy;
- the rent, or depreciation, rates and maintenance of these offices and other facilities such as workshops, stores and playing fields;
- heating and lighting;
- the office furniture and furnishings;
- the office equipment including IT provision, facsimile and telephones;
- the various insurances that need to be provided;
- company cars, fuel, personal car allowances and travel;
- postage, stationery and printing;
- the provision of subsidised canteen facilities and welfare in general;
- the costs of advertising and public relations;
- entertainment expenses;
- auditors and other outside consultants;
- interest to be paid on retentions and other loans.

The amount all these costs can be readily calculated to give an annual sum. This sum is paid for by adding the percentage to successful bids.

To give a simple example of how overheads might be charged (there are more sophisticated accountancy methods for this) assume the head office overheads are £100,000 p.a. and the annual turnover planned for in the company business plan is £2,000,000, then the percentage needed to be added to each contract to cover the overheads is:

$$\frac{£100,000 \times 100}{£2,000,000} = 5\%$$

This figure of 5 per cent assumes interest on retention money and working capital has already been taken care of when calculating the total office overhead.

When business is tight there is a temptation to reduce the percentage overhead to less than the amount calculated and to take zero profit to become more competitive, but this is a dangerous policy. What to do under these circumstances becomes a strategic decision, because the question has to be asked whether or not the economic situation of the time is going to continue and if so whether or not it is sensible to remain in this market place.

If the business is growing successfully and has outdone its budgeted turnover, a different situation arises. For example, if after say nine months the business obtained equates to the annual turnover planned for, it may be that the overheads already added to the work won has now covered the annual overhead bill. If so, then it would not be necessary to add any overhead to any other tenders submitted for the remaining part of the year, resulting in more competitive and successful bids. Alternatively, the company could add the same percentage of overhead or somewhere in between. This is a strategic decision. However, if the business grows too fast, this will mean recruiting more staff. They have to be trained and it takes time for them to absorb the culture of the organisation. Once they are 'full' members of the team, there would be some reluctance in parting with them, which means the work required the following year would have to be increased to keep them in employment.

Deciding on the level of profit to add is a senior management decision. It is a complex decision, as many factors have to be considered, but in the end it boils down to the instincts and experience of the management. This is after all why they are in their senior position and if they get it wrong to often, receive their marching orders. Examples of the issues upon which they make their decision include:

- How seriously do they wish to win the contract? This is influenced by
    - the strength of the current order book
    - the wish to break into a new market
    - the knowledge that this is the first of many contracts with this client
    - it is an existing client who they have built up a good working relationship with
    - it is a prestigious project which will raise the profile of the company.
- The company has worked with the client and or the design team before and all the contracts have run smoothly and profitably.
- What is the other competition for the work? Experience will indicate how successful bidding against these companies has been in the past and what the difference in margins has been.
- Will winning the contract overstretch the resources of the company and what are the implications of this?
- What levels of profit have been obtained on recent successful similar bids as this gives an indication of the market and where to pitch the bid? In other words, how much will the market stand?
- How much and how has finance to be provided to fund the earlier part of the project because of negative cash flow?
- Have changes to the standard form of contract been made so as to shift the risk away from the client and towards the contractor and if a non-standard contract what risks are involved?

- Are there production risks involved because of lack of information upon which the tender was priced, or are there complex and difficult construction details, or novel design features?
- How tight is the programme to complete the building and what is the likely risk of overrun and consequent penalty implications?
- Having reviewed all the above what is the overall risk if the bid is successful?

10.21   Bringing the tender information together for final review

The purpose of this report is to provide the senior management with all the relevant information needed to enable them to make an informed commercial decision on whether or not to proceed with the tender submission and if yes, then how much profit and overheads should be added. Some of the information will be produced in case it is required, but may not be used. The contents below inevitably overlap, but in general the information needed includes:

General contract information

- A brief description of project and its location to give a general flavour of what the contract is all about supported by appropriate drawings.
- Details of the client and consultants, which reminds senior management of past experiences if worked with before.
- The form of contract being used.
- Specifications, ground investigation reports and any other specialist reports.

Production information

- An outline description of the methodology of construction used as the basis for the estimate.
- The programme duration and confirmation this complies with the client's requirements.
- Any information, which would be pertinent to offering a reduction in the time of the construction programme.
- Any unresolved technical problems which might increase the financial risk.
- Any major assumptions made in the estimate, for example, permission to over fly with the tower crane had only been agreed in principal.
- Site visit report.

Contractual matters

- A list of special or unusual contract conditions imposed by the client and risks as this may effect judgement on profit figures.
- Special terms or conditions stated by any of the sub-contractors, especially those nominated.
- Any unresolved contractual problems that might increase financial risk.
- Any client's special requirements such as bonds and insurances.

Financial matters

- The overall cost of the project before overheads and profit are added
- A summary of the completed bills of quantities or tender summary. Management is not interested in the detail, just the final figure and a breakdown of this into discrete but usable information, e.g.:

  - own labour costs
  - materials cost and bulk quantities and materials comparisons forms
  - plant costs and comparisons
  - domestic sub-contractors' work costs and comparisons
  - nominated sub-contractors and suppliers
  - other provisional sums
  - overhead contribution based on company formula
  - schedules of suppliers, sub-contractors and plant and duration
  - schedules of prime cost sums and attendances.

A typical financial summary spreadsheet is shown in Table 10.19. It is designed so the estimator can immediately modify any changes management may wish to make at the final review meeting (section 10.22). The percentage analysis calculation column is based on the revised estimate figures, expressed as a percentage of the net total.

The adjustment column is based on the estimator's feel for the commercial market and how much they believe the quotations received from others could be renegotiated if the company wins the contract. The percentage columns are useful in the final review meeting as an indicator of the proportion of the work set against the total price before overheads and profit are added. This gives information to management to assess the risk and decide whether the proportions are consistent with previous successful bids. The amount of suggested overheads and profit could be entered here to give management an indicative price, but may be omitted and entered at the final review meeting. The dayworks and attendances are separated in this case to permit management to make a separate decision on the amount of profit and overheads to add, which may differ from the main part of the tender.

Table 10.19   Tender summary sheet

Tender summary

Contract: Jasper Winston Offices

Description		Estimate (£s)	Adjustments	Revised estimate	% age analysis
Own work	Labour	103,542		103,542	7.12
	Plant	25,678	−1.0%	23,110	1.60
	Materials	323,456	−1.5 %	318,604	22.09
Domestic subs	Net	784,230	−2.5%	764,624	53.02
Nominated subs	Net	25,600		25,600	1.78
Nominated suppliers	Net	3,456		3,456	0.24
Total		1,265,962		1,238,926	85.85
Provisional sums					
Preliminaries					
Employer's requirements	Insurances, etc.	12,659		12,659	0.88
Contractor's requirements	Staff	48,432		48,432	3.37
	Accommodation	5,600		5,600	0.39
	Services and facilities	8,098	−0.5%	8,056	0.58
	Plant	134,654	−1.0%	121,189	8.41
	Temporary works	6,800		6,800	0.47
Statutory authorities	Water-connect	380		380	0.03
	Electricity-connect	250		250	0.02
Sub Total		214,280		203,366	14.15
Net Total		1,480,242		1,442,292	
Overheads	5%	74,121		72,115	
Profit	2.5% pre-overheads	37,061		36,057	
Prime costs/ attendances		7,000		7,000	
Estimated dayworks		20,000		20,000	
Tender value		1,618,424		1,577,464	

Estimator's comments

- The quality of information on which the estimate was made and any assumptions as a result.
- The estimator's view on the current market place and conditions.
- The estimator's view on the likely profitability of the contract that is often a function of how buildable the works are.
- Who are the other competitors?
- The likelihood of more work from the client especially if the one under review is the first phase.
- Recommendations of any conditions the contractor should lay down when submitting the tender.
- Date and time of submission.

10.22 Final review meeting

This meeting is to make the final decision on the tender submission price, which includes the overheads and profit. The key players in this are at the very least the following:

- The senior manager, usually the managing director, who has the overall picture of the financial state of the business and is commercially aware of what is in the company's best interests.
- The estimator who completed the process and is the most familiar with the tender documentation and price.
- The buyer, or in some organisations the QS, who is there because of their commercial awareness of the current marketplace; very important when considering the possibility of reducing some of the estimator's prices for sub-contractors and suppliers.
- The planner whose particular contribution is concerned with answering resource questions and the implications of altering the sequence of work or reducing the overall time scale.
- It is an advantage to have the contracts manager there. Failing that, it is important that a 'practical' input is available to discuss methodologies in the event of changes being made and to cast an eye over the chosen method used for the basis of the tender.

Most companies have a standard agenda for this meeting, reflecting the topics discussed in section 10.21. In the background is the overarching consideration of what the client will use as their criteria of selection and could be dependant on the nature of their business. Typical issues might be as follows:

- Is price the overwhelming consideration? It is more likely to be if a public body and may be in their constitution as a standing order. There may be some history of the client that demonstrates price is not always the key criterion.
- The client's business may well indicate the sensitivity for completion on time or any advantages gained by an accelerated programme, even though the latter may cost more. Examples would include retail developments when completion is keyed into the Christmas selling and sports stadiums being built primarily in the closed season. Any acceleration, which limits the loss of sales, could be advantageous.
- Whether or not the client is experienced can affect how much effort and time is spent on presenting such information as methods and issues of environmental, quality and safety.
- Equally if the client has a particular interest, say environment, it would be appropriate to design the contents of the final document and/or presentation appropriately.
- If an oral presentation is to be made, the contractor needs to consider the format and content prior to putting the information together for rehearsal.
- Finally, the client will always be interested in the amount of money that has to be spent on the project. On the one hand this could be presented as a total sum or alternatively as a predicted monthly outlay. However, the latter may be seen as overstepping the mark and cause irritation. Equally, excessive front-loading (10.23) or apparent high fixed costs in the preliminaries could have a similar effect.

The purpose of the meeting is:

- To make the final decision as to whether to submit the tender. Normally this decision would have been made earlier on receipt of the tender documents (10.4), but issues may have arisen subsequently and this would be resolved at this meeting.
- Whilst the tender will have been produced based upon a selected method of construction, management needs to be assured that this is the best way, so alternatives need to be evaluated. Another method may be in the client's interest if the effect is to be able to complete the contract earlier.
- To reduce (usually) the prices of the domestic sub-contractors and suppliers. The reason for the opportunity to reduce prices is based on the 'bird in the hand' principle. Being asked for a quotation is very different from being awarded the contract. The estimator and especially the buyer

have a good commercial feel for what the market will bear. Based upon current trends, the amount of work available, and the relationships with the suppliers and sub-contractors, it is not unusual to be able to reduce the sums between 1 and 3 per cent. An indication of possible reductions may have already have been indicated by the estimator as shown in Table 10.19, but the buyer will be even more sensitive to the current market place and is able to fine tune these figures.

- Any other modifications to the contract are considered at the meeting, especially if the client has deleted clauses which transfer more risk to the contractor
- As a result of the deliberations at this meeting, the contractor may decide they can offer an improved service than that which the client asked for. This can include, improved functional performance of the building, reduced costs by modifying the design and a shorter construction programme.
- The estimator has in the example shown in Table 10.19 added a figure for overheads and profit. This meeting gives management the opportunity to reassess both in the light of the information brought to their attention and the considerations outlined in section 10.20.
- The final decision on the costs has to be made. The advantage of the use of computer-aided software is that the modifications requested by management can be entered at the meeting and an instant answer produced. This gives management the opportunity to try out a variety of different options to assist in making the final decision.

10.23 Final adjustments and submitting the tender documents

As a result of the review meeting the tender documents are modified accordingly. It may not be desirable to show the full extent of overheads and profits. As bills are usually priced using computer software, it is not difficult to add a percentage to all bill items thereby reducing or eliminating the shown profit figures. This also can be carried out for a proportion of the preliminary items and in the quotations given by the domestic sub-contractors. However if this is done too much it will become clear to the client's QS that the bill items costed do not reflect the marketplace.

There are a variety of ways the bills can be modified to benefit the contractor financially, but at the same time not increase the overall tender price. When the contract commences there is a period of approximately two months before the contractor receives money from the client for work done.

This, coupled with retentions, causes a negative cash flow during the early part of the contract resulting in the contractor having to financially support the project at the beginning. To compensate for this the contractor sometimes uses a technique called 'front-loading'. This is where the estimator inflates some of the larger items at the beginning of the bill and to compensate, reduces some of those which will be carried out towards the end of the contract. Whilst there is a general acceptance that this occurs, there is a limitation to how much this can be done before the client's QS would raise questions. If the contractor becomes insolvent during the contract and there has been excessive front-loading, then the client would have to pay out more to have the rest of the contract completed by another contractor. Unless site management is aware of this strategy and the amount of the adjustment, they can be lulled into a false sense of security believing they are profitable at the beginning of the contract, when they are not. The practice does raise some ethical questions.

The opposite of front-loading is that of 'back-loading' which is only relevant in times of high inflation and where fluctuation clauses are included. Here items to be completed at the end of the contract are inflated at the expense of the earlier ones resulting in the contractor being paid proportionately more: although this is a dubious practice.

Another technique used is identify certain items in the bills that currently are shown as small quantities, but it is suspected are either a measurement error, or are likely to increase as a result of variations, sometimes referred to as 'item spotting'. If the price for these items is inflated, the effect on the overall tender price is minimal, but if the measure is subsequently increased then the contractor can make substantial gains. An example of this would be in the excavation of foundations when unexpected problems are revealed, such as pockets of peat, and much more has to excavated and carted away as a result. However, to do this conflicts with the current philosophy of openness between all parties and so has much wider implications.

One cannot underestimate the need to submit the final documents in a well-polished form. Presentation has become increasingly important over the last few years and reflects the image of the company. If an oral presentation is required to support the bid, then all the participants have to be well coached and rehearsed (*Business Organisation for Construction*, Chapter 11). The content of the accompanying letter, which will outline any modifications or suggestions, needs to be carefully composed to communicate clearly, but written in such a way as not to cause offence.

It is normal for the documents to be sent in a sealed envelope, known as a sealed bid, so they can be kept under lock and key until all are opened at the same time under strictly controlled conditions to ensure honesty prevails.

It is not unusual for the bid to be delivered by hand either by courier or a company representative. Sometimes this is because the contractor has been on a very tight schedule, or because it leaves an opportunity to make last minute adjustments if new information arises.

After the results of the bids have been announced, and the successful bidder notified, it is increasingly common practice for the client to notify all competitors of each other's bids. The timing of the disclosure of this information must remain confidential until after the contract has been signed otherwise the lowest bidder could try to negotiate higher prices. This needs only to be a list of the various tender prices, as some contractors may not wish their competitors to know what their bid was. If this is not published, then it is important for the estimator's grapevine to come into play to discover the value of the other bids. This applies whether successful or not as these data give an indication of the current marketplace and the company's relative position in it, which is useful knowledge for current and future tenders (Chapter 11).

10.24 Vetting of tenders

The final stage in this process is with the client. They will usually open all the tenders together and list the contractors' bid prices. This is done in a controlled manner by the client, or representatives, and would normally include the quantity surveyors responsible for putting together the bills and contract documents. It should be done as soon as possible after the deadline given for submission of the bids. Precisely how this panel is constituted is not prescriptive. What matters is that it is, and is seen to be, a fair and honourable process. It is normal for two members of the panel to countersign each tender submission. Any unsolicited bids must be rejected and any late bids returned to the sender unopened.

All except the lowest three should be advised immediately their bid has been unsuccessful. The documents of the lowest of these remaining bids are then vetted and checked for genuine errors and confusion. The most common occurrences are:

- mathematical errors made when adding up the columns of numbers, decimal point errors and multiplying quantities with rates;
- obvious errors such as pricing a material in say kilograms rather than tonnes, or clearly pricing for a lower or higher specification than is required;
- prices for similar items at different parts of the bills, which are considerably different in price;

- extreme cases of 'front-loading', 'back-loading' and 'item spotting' – in all of these cases, the contractor cannot be compelled to change them, but the QS can advise the client not to accept the bid.

The contractor should be advised of these errors and given the opportunity to revise its bid. The contractor then needs to consider whether or not it wishes the revised bid to remain on the table or withdraw it. It could be that these amendments raise the tender price above the next lowest, in which case this one is subjected to the same rigorous inspection and so on until a final decision can be made as to which one is selected to execute the contract.

If the lowest tender exceeds the client's budget price and a maximum sum has been indicated which must not be exceeded, negotiating will take place. If accepted the contractor would usually accept the risk with the following caveat, 'the price is for the scheme as drawn now'. There may well be other clauses and exclusions included to protect the contractor. They will try to spread the risk to the suppliers and sub-contractors. For example if the price calculated by the contractor was 5 per cent above the guaranteed maximum price, the sub-contractors would be asked to reduce their price by a similar margin.

In the event that the client had valued the job at say £42 million and the contractor believes the cost to be £44 million, the contract might be let at £44 million on the condition that anything saved between £42 and £44 million would be handed back to the client and anything below £42 million would be divided say 25 per cent to the client and 75 per cent to the contractor.

10.25 Risk and uncertainty in estimating

Construction is by its very nature an industry with a significant amount of risk. The estimating process carries its fair share and whilst not intending to become involved in the mathematical and statistical analysis of risk evaluation, the reader should be made aware of the issues involved. This is a summary of the issues the estimator takes into account when pricing a contract and management when reviewing the tender prior to submission. The factors built in by either the client or designer, which can impact on the estimating process, are:

- The accuracy and availability of the tender documentation necessitating making assumptions that may have inherent error.
- A very tight project programme is likely to overrun as most programmes have some optimism built in and any delays, which occur, are more difficult to reclaim. This can result in penalty clauses being invoked.

- Fixed-price contracts have to take into account most increases that may occur in the future. The longer the duration of the contract, the more difficult these are to predict thereby increasing the financial risk.
- Prime cost sums are on the one hand an advantage in so far as the client takes the financial risk, however, the contractor has less control on performance that may have an impact on the overall performance of the project.
- Complex projects are more difficult to price and quantify in terms of duration.
- Monthly interim valuations have to be agreed by the client and contractor. Over-zealous quantity surveyors representing the client's interest have a tendency to try to undervalue the work to be seen to be protecting the client's interests. Even if the valuation is only a small percentage lower than the true value it will have a knock-on effect on the contractor's cash flow.
- Any variations to the works are measured separately. If it is appropriate to use the bills rates then there is no inherent risk. Unfortunately it is not as simple as this and there is a risk attached especially if a large number of variations are instructed.
- The way architects, engineers and the clerk of works interpret the specification can cause problems, which can be costly and cause delays. For example a contractor, with the agreement of the consultant engineer, decided to precast the reinforced concrete stairs on site in lieu of in situ to permit ready access between floors during the construction. The precast stairs were rejected initially on the grounds that the blowholes formed in the surface, whilst complying with the in situ specification, did not meet that specified for precast.

The following risks are inherent in the tendering process itself:

- Whilst taking account of the amount of time predicted to be lost as a result of inclement weather, it is notoriously difficult to assess.
- In the euphoria of wanting to successfully obtain the contract and meet the client's needs there is a danger of making optimistic assumptions when programming.
- The time available for carrying out the estimate is limited so there is always the risk that not enough time is allocated to thinking through all the implications of the construction processes and the wrong method of working is selected as the basis for the tender.
- Deciding upon the levels of overheads and profit to be added is the greatest risk of all, which is why senior management takes the decision.

- Predicting the increased costs of plant, materials and labour throughout the duration of the project, especially if a long contract is difficult. The risk is reduced in the event of having a fluctuation price contract.
- If during the course of the project shortages of labour, plant and materials occur, the price of obtaining them will rise.
- In the end, other than when insufficient overheads and profits have been added, the contracts success or failure is usually a reflection of the quality of the management.

It is interesting that many of the risks can be reduced significantly if all parties work with a spirit of cooperation rather than confrontation.

10.26 Computerised estimating

Computer-aided estimating has come of age and is accepted by most medium and large contractors. Software packages are also available for the smaller organisations. The process described in this section covers traditional methods of procurement where bills of quantities are still in use. However, the general principles apply to all procurement where both detailed and estimates based on approximate quantities are used.

10.26.1 Entering the bills of quantities

The first stage of the process is to enter the bills of quantities into the system. This can be executed in three ways:

- Re-typing the bill into the system, but this is time consuming, tedious and prone to errors.
- Putting the bill of quantities through optical scanners to prepare an electronic bill of quantities. Depending upon the type used there can be a need for checking as errors occur as some types of scanners misinterpret the information.
- Using an electronic version provided by the client's QS. This saves time providing the contractor's and quantity surveyor's software are compatible. Most of the technical problems have now been overcome, but the use of them is a two-way process and both receiver and sender have to understand the needs and methods of working of each other. If the contractors do not ask for them, there is little point in the quantity surveyors producing them. There is however increasing use of this method and it is preferred by many.

10.26.2 Linking the bills to the estimating library or database

The quality and quantity of data and the speed of access are the keys to the success of any computer-aided system. A typical structure is a resource-based model in which data are primarily divided and stored in two categories: that which changes and needs regular updating, such as the costs of labour, materials and plant; and that which is relatively static, such as production rates of output. This type of system is most appropriate for unit rate estimating.

There are two fundamental ways of relating the bills to the data stored in the estimating library. First by storing the data in the library so that it matches typical bills items. They may not always match perfectly so some manual adjustments may be necessary. Second to collect and keep the most up-to-date prices for the more frequently used resources. These would include all-in labour rates, gang sizes and outputs, different mixes for concrete and mortar, typical plant costs and outputs, and materials costs. Although the library will store many inclusive rates, the estimator has to build up more bill item rates from the basic data stored.

In both cases, when developing a library, the estimator needs to be assured that the computer-aided estimating system is capable of supporting the approach to be taken, and before entering the data, is fully cognisant of the way it works otherwise the full benefits will not be realised.

It is possible to purchase a library of data with most systems and then modify or adapt the data according to the contractor's view on costs and production. The advantage of these packages is they provide a ready filing system which estimators can adopt without designing their own. However, before purchasing such data it is important to investigate the package to ensure the ways the information is calculated coincides with the company's. Databases are also available on the Internet.

Any system must have the ability to include and calculate variables that are either pertinent to the contract or the company experiences. Typical examples of these include:

- allowances for waste on materials that vary depending upon material, the complexity of the contract and the history of the contractor in controlling waste;
- bulking in aggregates, especially sand as a result of excess water;
- conversion factors; for example, aggregates and sand are purchased by the tonne and are measured in m^3 or m^2, bricks are obtained in 1000s and normally measured in m^2

- materials suppliers offer different discounts often with conditions attached – the system needs to be able to deduct these from the calculation;
- having sub-systems that permit analysis of the quotations from suppliers and sub-contractors;
- having mechanisms that identify all outstanding quotations from suppliers and sub-contractors;
- having cost codes which are being used throughout the company;
- for contractors working overseas it would be useful for monetary conversions;
- some companies want the mark-ups to be added to each individual bill item rather than as a percentage at the end;
- if the contract is fixed price, but of such duration that inflationary factors are likely to come into play, an assessment of the likely increases has to be made and added.

The estimating library is continually modified from various sources, namely:

- performance data, such as plant outputs in different conditions and labour rates for executing different tasks with a variety of materials of prescribed dimensions;
- costs which require regular updating such as labour rates, CITB levy and NIC;
- material suppliers costs;
- sub-contractors costs.

The estimating library can then be used to provide information for producing the preliminaries, bill item rates, or approximate quantities rates. The estimate can then be complied from the suppliers' and sub-contractor's quotations, the bill build-up rates, and the preliminaries to provide the document required for the final review meeting (section 10.22).

At the final review meeting, any modifications the management require can be entered and immediate feedback given on the effects of these decisions, which gives the opportunity to inspect various scenarios before making the final decision on the tender price.

References

Aqua Group (1999) *Tenders and Contracting for Building*, 3rd edn, Blackwell Science.

Brook, M. (1998) *Estimating and Tendering for Construction Work*, 2nd edn, Butterworth Heinemann.

Buchan R. D., Fleming, E. and Grant, F.E.K. (2003) *Estimating for Builders and Surveyors*, 2nd edn, Butterworth Heinemann.

Chartered Institute of Building (1997) *Code of Estimating Practice*, 6th edn, Blackwell.

Co-ordinating Committee for Project Information (1987) *Co-ordinated Project Information for Building Works*, Co-ordinating Committee for Project Information

Griffiths Complete Building Price Book, Glenigan Cost Information Services

Harris, F. and McCaffer, R. (2001) *Modern Construction Management*, 5th edn, Blackwell Science

JCT05 Intermediate Building Contract (2005), RIBA Publications.

JCT05 Standard Form of Building Contract with Quantities (2005) RIBA Publications

Latham, M. (1994) *Constructing the Team*, HMSO.

Laxton's Building Price Book, Butterworth-Heinemann.

Sher, W. (1996) *Computer-aided Estimating: A Guide to Good Practice*, Longman.

Skoyles E.R. and Skoyles, J.R.(1987) *Waste Prevention on Site*, Mitchell.

Smith, A.J. (1995) *Estimating, Tendering and Bidding for Construction Work*, Macmillan Press.

Smith, R.C. (1999) *Estimating and Tendering for Building Work*, Longman.

Spon's Architects' and Builders' Price Book, E&FN Spon

Standard Method of Measurement of Building Works (1997), 7th edn, RICS.

Walker I. and Wilkie, R. (2002) *Commercial Management in Construction*, Blackwell Science.

11

Bidding strategy

11.1 Introduction

In capitalist society most business is obtained by tendering a price to carry out work or sell a product or service in competition with others, the lowest being the winner. An alternative approach is to auction off an item, starting with an initial figure and allowing competitors to outbid each other until only one is left paying the highest price. The flip side of this is the concept of bargaining when a price is offered and the purchaser sets about negotiating the proposed price downwards. In all of these cases the purchaser and/or the seller needs a bidding strategy for the process. In essence they need to establish a range from the lowest to the highest price that is acceptable to them. Bidding is the relationship between the levels of profit and overheads required, the accuracy of the estimate and the number of competitors.

It is necessary to develop a bidding strategy because of the uncertainty associated with tendering for work. This is made more difficult because of the low mark-ups added by the contractor, resulting in an increased risk of winning the contract and losing money at the same time. Ideally the contractor wishes to produce the highest possible tender price that will obtain the work, but which will also cover overheads and contribute towards profits. To achieve this the contractor relies heavily on historical data provided from previous bids and analysis of the competitors.

The purposes of any bidding strategy are :

- determine whether to tender for the work as there is no point in wasting time and effort pursuing an already lost cause;
- increase the chances of winning the contract(s);
- obtain an adequate return relevant to the risk taken;
- maximise the profits;
- if bidding for as a losts leader, minimise expected losses;

- minimise losses, if the marketplace is very tight and too low a price has been submitted;
- provide an even workload throughout the financial year.

Before looking at procedures of producing a strategy it should be noted there is still considerable scepticism by much of the UK construction industry in using a mathematically based system in lieu of working from instincts and experience. Also many of the models produced do not allow for these subjective judgements to be included. Indeed, some would argue that the significant losses made by companies in recent years have not been as a result of bidding, but because of the errors made during the estimating process in misunderstanding the implications of the problems of the construction work. If the estimating quotation is not accurate, with margins being as low as they currently are, the proposition that strategic bidding is of value collapses.

So why should bidding models be used since the industry has apparently successfully developed without them? It is argued that it is not so much whether they should be used, but rather what are the principles behind them, which will clarify the thinking processes of those making the decisions and by doing so supplement and improve their subjective judgements. It would be imprudent to suggest that substituting a mathematical model should ignore the expertise and commercial acumen gained. Currently, these models are more likely to be found in the research arena of academia than in the boardroom. However there will be a divergence as the models are seen to be relevant and the estimating process becomes more accurate or margins increase.

11.2 Comparisons with competitors

In 1956 L. Friedman originated one of the most accepted methods of comparing competitors. The process is relatively straightforward, but relies on having competed against the same contractor on several occasions. Each of the specific competitor's bids is expressed as a percentage compared against the estimated cost (not the price tendered by one's own company), i.e. net before overheads and profits are added. This information is then presented as a frequency distribution or a histogram form as shown in Figure 11.1. In this example competitor A has bid 27 times against the company. For the purposes of this demonstration the bids have been grouped in bands of 2.5 per cent. There could be situations where there is a negative showing on the histogram if the competitor's estimated cost is lower than the company's.

Since the comparison of the competitor's bid is the total tender price compared with the net cost, it can be seen, for example, that if the company

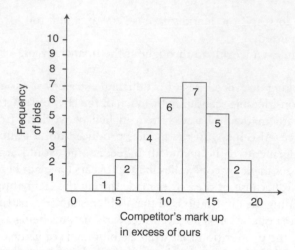

Figure 11.1   Frequency distribution of competitor A

had added 5 per cent to its estimate, it would have lost 3 of the contracts (11.1%) and won 24 (88.9%), and if 10 per cent had been added then 13 (48.1%) would have failed and 14 won (51.9%). The probability of winning the work in these two cases is 0.89 and 0.52, respectively.

Using the data shown in Figure 11.1, a cumulative frequency curve, Figure 11.2, can be plotted showing on the vertical axis the probability of beating the competitor and on the horizontal axis the percentage mark-up. If zero mark-up is added then the probability the company will win the work is 100 per cent, and if 20 per cent is added the probability of beating competitor A is zero. A word of caution though, this form of analysis relies on having competed enough times to have sufficient number of comparisons to be statistically valid.

The same process can be adopted for all competitors added together to assess where one fits into the market. However, alternative models have been proposed since Friedman and there has been considerable debate as to which is the most accurate. Interested readers might read Chapter 5 of Harris and McCaffer (2001) for a concise and pertinent review of these.

11.3 The accuracy of the estimate

Much of the validity of this approach depends on the accuracy of tenders because of low mark-ups. To apply such techniques requires the estimator to know his or her range of accuracy so that comparative information can be logically applied. Different estimators in the company and in competitive

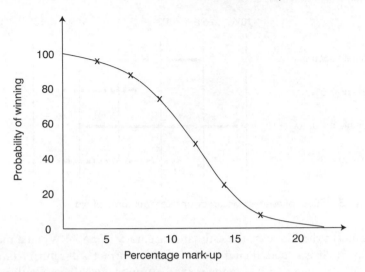

Figure 11.2   Probability curve of outbidding competitor A

companies will vary to a certain extent, but generally they aim to produce the real cost of executing the works.

As indicated in Chapter 10, the estimator must make various assumptions when preparing the estimate, all of which have a margin of error associated. These include such issues as:

- That the all-in labour calculation reflects reality (Table 10.5) when variable assumptions were made about inclement weather, the bonus to be earned, overtime worked and levels of sickness.
- The selection of the productivity rate to be used in calculating the bills rate (section 10.10).
- The percentage waste allowed for materials (section 10.9).
- That the programme will be adhered to, which will have an impact on the preliminaries.
- Inflation costs on materials labour and plant.

The result of these assumptions is that the estimator will produce a likely cost estimate range of the true costs of say $\pm N\%$, the value of N depending on the accuracy of the estimator. If the senior management assumes the estimate cost is correct and adds a percentage to cover for overheads and profit, then one of two outcomes can be expected depending on which end of the range is the estimated cost. If at $-N$ then the contract will probably be won making either a loss or minimum contribution to overheads, and if $+N$, the tender price will be too high and bid unsuccessful. Figure 11.3 illustrates

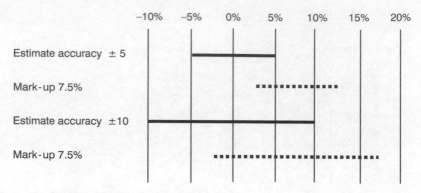

Figure 11.3   Effect of estimate accuracy on likely outcome of bid

this point. In the first case the estimator accuracy is ±5%. When a mark-up figure of 7.5% is added, then if successful the contract will return from 2.5% to 12.5%. In the second case where the estimator accuracy is only ±10%, the return will range from between −2.5% and +17.5%.

Studies carried out using simulation programmes have indicated that if the estimator estimates more accurately, then the achieved profit margin will be increased, which confirms the belief that the more confident senior management is about the estimate, the more likely they are to pitch their mark-up at the right level, especially if they take into account knowledge of their position in the market and against the competitors for the tendered project.

More accuracy can only be obtained if estimators have accurate data at their disposal. This emphasises the need to keep up-to-date records and to have appropriate feedback from the site of the actual costs, causes of delays, etc. This can only occur if the site has a good post-contract cost-control system in place (Chapter 13) during the construction phase and from feedback after the final account has been settled.

11.4 The number of competitors

What is clear is that the more competition, the less likely the odds are on being successful. It is therefore in the interest of the contractor to select contracts with a limited number of competitors. The reason for this is that research suggests that compared to the mean bid of all the competitors, the lowest bid reduces as there are more bidders, perhaps due to stiffer competition.

11.5 Summary

There are steps that can be taken to develop a bidding strategy. First, establish the base data by:

- investigating all previous contracts tendered for and tabulate the names of the competitors on each contract and their tender price including one's own company;
- producing a comparison of each major competitor's price against one's own;
- constructing a probability curve for each competitor;
- constructing a probability curve for all major competitors;
- updating both probability curves after the result of every tender submission.

Then, second for the specific bid:

- identify the competition;
- construct a graph superimposing all the competitors' probability curves and determine the probability of being successful based on this information;
- make a judgement as to what mark-up should be added to stand a good chance of being successful with the bid, or if this is insufficient to satisfy the company requirements either decline the tender or submit the tender with the mark-up needed to remain profitable and in business.

Finally, it should be noted that the marketplace is changing. Many contracts are now not awarded just on price, but rely on negotiation when the contractor is selected on other issues, such as design, reputation, previous work for the client, meeting deadlines, and so on.

References

Cook, A.E. (1991) *Construction Tendering: Theory and Practice*, Batsford.

Friedman, L.A. (1956) Competitive bidding strategy, *Operations Research*, 4, 104–112.

Harris, F. and McCaffer, R. (2001) *Modern Construction Management*, 5th edn, Blackwell Science.

Park, W.R. and Chapin, W.B. (1992) *Construction Bidding: Pricing for Profit*, 2nd edn, Wiley.

Purchasing

12.1 Introduction

It is not intended to give a detailed analysis of the purchasing function as much has been considered in other chapters such as in Chapter 10, but rather to address some of the key issues affecting this role in an organisation. Purchasing is sometimes referred to as procurement, but this can be confusing in the construction industry; procurement is also applied to the process of selecting a contractor to carry out the work. It is also referred to as the buying function.

12.2 Ethics

It is important to put the subject in context and consider the ethical implications of purchasing. Morality and ethics affects every branch of the organisation, but the purchaser in a construction company has high spending power. There has been increasing concern in recent years and various organisations, notably the Co-operative Bank, have been developing ethical business practices. There are two main considerations: exploitation of labour, especially in developing countries; and environmental concerns. Both have serious implications for sustainability summarised in the Natural Step's fourth system condition 'that in society people are not subject to conditions that systematically undermine their capacity to meet their needs'.

Typical examples of the problems the purchaser should consider include:

Exploitation of labour in developing nations

Labour here is consistently paid a low wage, which does not permit workers sufficient finances on which to live satisfactorily, or surplus money to invest in their business, such as agriculture, or to purchase material goods. In the case of agriculture, if farmers are able to invest in their businesses by, for example, drilling a water well to yield better crops, then they generate more finance which in turn allows them to purchase goods. Once labour is able to purchase goods, then somebody needs to manufacture them. The result is improved standards of living and a more sustainable and equitable society. That is a simplistic view on very complex issues but was highlighted in the Independent Commission on International Development Issues publication *North–South*. Only by raising the standards of a given country can that country begin to make strides towards development. This is a very complex issue, as whilst it is naturally considered to be unacceptable by those in the developed world, if a society is poor it cannot afford education for its children, so what should the children do instead? Only by raising the standard of that country can education be made more widely available to children. The greater the finance available the higher the age to which education can be offered.

Exploitation of labour in developed countries

It is often forgotten that exploitation of labour still happens in the developed world. The employment of illegal immigrants is an obvious example, but many people, especially women are exploited by being paid to carry out work in their home at rates of pay well below the national minimum wage usually paid on piece work rates, i.e. based upon output rather than an hourly wage. This is difficult for a purchaser to identify, because it is hidden from view.

Exploitation of the environment in the manufacture of materials

There are relatively strict regulations in the developed world concerned with environmental pollution during extraction and manufacture of materials and components. However, many materials are sourced from elsewhere in the world, where legislation either does not exist or is less stringent. In effect, this means Western countries can export their environmental problems. Typical examples are obtaining timber from non-renewable sources and the extreme amounts of uncontrolled pollution caused in the extraction of metals such as waste rock and slurries, not to mention the pollution of watercourses. There is much documentation of these occurrences, but the purchaser may have some difficulty ensuring the material bought actually comes from a less polluting source. Timber from sustainable sources has over the years become easier to trace.

Unnecessary transportation

Linked to environmental issues is the unnecessary transport of materials. As a simple example, a design decision specifying Skye marble extracted approximately 150 miles north of Glasgow was sent south by road to Manchester over 350 miles away from the quary to be fixed on the face of precast concrete panels then transported back to Glasgow to a building being constructed there.

Unscrupulous suppliers

There are recorded examples of suppliers and those in the service industry who have sufficient financial muscle to lower their prices for a period long enough to put the competition out of business enabling them to monopolise the market. Others have been accused of stealing ideas from others and bringing them to market first.

Gifts and favours

This is a highly contentious issue, but interestingly what is acceptable varies from generation to generation, company to company, and society to society. It is easy to take the moral high ground and state categorically that all gifts and favours are a bribe so none should be accepted, but it is argued here that this is a decision either the senior management makes on behalf of the company or one an individual takes. A late morning meeting overruns and someone suggests continuing over lunch: who should pay? Christmas presents vary from a diary, calendars and bottles or crates of wine or spirits. Firms have hospitality suites at important sporting or entertainment venues which are offered as a 'thank you' for the business received. In extreme cases an all expenses paid holiday for the entire family can be offered. At what point do these become a bribe? It is a problem because many of the smaller gifts especially those offered at Christmas have become a tradition and are generally accepted. The question left to the reader is what would you feel about a company if they had regularly given you a bottle of whisky and then for no reason it does not arrive this year?

Conflicts of interest

These are such situations were the purchaser is buying products from a relation, a company in which they own shares or from close friends. The transactions may be honourable and honest and in the best interest of the company, but can raise suspicion from others. In these cases, it may be sensible to keep such suppliers, but the deals need to transparent and clearly vetted from senior members of the organisation, so everybody can see the deals are above board.

Tax evasion by supplier or sub-contractor

In spite of continuing efforts by the UK Treasury, there is still a significant number of people, many of whom are employed in the construction industry, who are either evading the payment of taxes or who are also being paid social security benefits of one sort or another. This is often referred to as the black economy. It is only right and proper that a company using these services has systems in place to deny access to employment from these individuals. Equally, if a supplier is offering materials for cash, thereby avoiding the payment of VAT, this should also be resisted.

Unsustainable business practices

Prices, which are clearly well below the current market value, will almost certainly be because of one or a combination of any of the above practices. Besides having an ethical dimension they could potentially fail to deliver resulting in delays and higher expenditure.

12.3 Fair price

With these issues as a background to the purchasing process, obtaining goods and services is not just about obtaining the lowest price. Whilst this is important, it is equally imperative a fair price is sought. This is because without a fair price the supplier cannot run a sustainable business and will either go out of business or have to take short cuts to survive.

A fair price must permit the supplier to:

- not be forced to carry out any of the issues identified in section 12.2;
- have sufficient income to pay fair wages, which improves the likelihood of maintaining a stable labour force;
- be able to invest in new plant and machinery and keep abreast with modern developments; this makes the organisation more competitive and enables the business to survive in the competitive marketplace;
- depending on the nature of the business be able to invest in research and development to ensure their products keep up with development;
- invest in people in their organisation; this involves development and training so personnel are up to date with appropriate legislation, new developments and management skills;

- use materials of the appropriate quality to the specification requested;
- produce a safe working environment.

12.4 Procedures for procuring suppliers and sub-contractors

This depends upon the size and scale of the procurement. If a sub-contractor is employed to erect a steel frame, the contractual arrangements and documentation would be very different from obtaining a plumber to repair lead around the junction between a domestic chimney and a pitched roof.

However, the purchasing process will involve some or all of the following stages:

- identifying the requirements at the estimating stage;
- determining the specification requirements;
- determining the delivery requirements;
- obtaining quotations from various potential suppliers some of which may be preferred suppliers;
- obtaining outline design proposals;
- receiving and evaluating the quotation;
- on winning the contract, re-checking the specification, delivery needs and re-negotiating the prices;
- issuing contractual documentation;
- tracking the supply to ensure the product or sub-contractor arrives as required; this may mean making visits to the factory to assess progress;
- monitoring performance.

It is important to understand, especially in the case of suppliers of materials and components, how their manufacturing process works. Failure to understand leaves one vulnerable to being misled or unable to sensibly negotiate prices and variations to the order.

12.5 Preferred suppliers and sub-contractors

It makes sense to engage suppliers and sub-contractors who on past record and performance are reliable. Hence monitoring of performance is crucial. It is important to consider what percentage of the supplier's business the order is. Too high a percentage and problems can occur as they will lack the flexibility to redirect production from another client to you.

New blood should be introduced from time to time, otherwise there is a danger of complacency and the possibility of missing new opportunities for development. This can be done from time to time, partially as an experiment, but not before the new organisation has been carefully scrutinised. This is carried out in a systematic manner, by interviewing and reviewing past performance with other organisations. In all cases it is important to set down performance standards against which the organisation can be compared and monitored. For example (although in some cases setting absolute standards would be difficult):

- delivery on time
- quality of material/workmanship
- response if things go wrong
- general co-operation with the main contractor and other sub-contractors
- attention to safety issues and standards agreed
- attitude to claims.

12.6 Supply chain management

This subject is explored in more depth in *Operations Management for Construction*, Chapter 7. For the purpose of these comments, the design stage of suppliers or services, such as mechanical and electrical, is ignored.

Supply chain management is a natural progression from having preferred suppliers and sub-contractors. Its success relies on a relationship of trust between parties. A partnership like this means a supplier needs to fully understand the requirements of the contractor and vice-versa.

The smaller supplier may not have the wherewithal to provide the necessary controls for such issues as total quality management (*Operations Management for Construction*, Chapter 8) the contractor needs, so the contractor has to assist them in this. They may also need financial investment to purchase the most up-to-date machine tools and other equipment. It will usually mean guaranteeing minimum quantities of orders each year so they can invest. The supplier/sub-contractor may also need to amend its internal controls systems to be compatible with the main contractor.

References

The Co-operative Bank www.cooperativebank.co.uk/ethics/ethicalpolicy (accessed 12 January 2009).

Independent Commission on International Development Issues (1980) *North–South: A Programme for Survival*, Pan Books.

Post-contract cost control

13.1 Introduction

For the purposes of this text, post-contract cost control is defined as the control of costs after the contract has been awarded to carry out the building work. The emphasis is on the cost control from the contractor's perspective, but the client, the sub-contractors and suppliers are also looking to control their costs. Mention will be made from time to time about the client's cost-control mechanisms, as whilst similar, the emphasis can be different especially since the client is usually looking at much longer period for its returns.

13.2 Pareto's Law

Pareto's Law states that:

> In any series of elements to be controlled, a selected small fraction of items in terms of number of elements almost always accounts for a large fraction in terms of effect.

Conversely the majority of items are only of minor significance in their overall effect. Simply put, 20 per cent of your decisions will affect 80 per cent of your business. This is similar in principle to 'management by exception'. A typical example to demonstrate this is approximately 80 per cent of all construction work is carried out by only 20 per cent of the companies, the remaining 80 per cent carrying out 20 per cent of the work. In other words it is unproductive to spend an equal amount of time controlling each element, in this case costs, but rather to concentrate on controlling the key elements. Figure 13.1 illustrates this point.

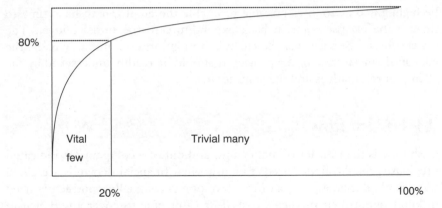

Figure 13.1 Pareto's Law

The significance in controlling construction costs is that there are potentially a large number of cost centres that can be created. The numbers of items in bills of quantities clearly demonstrates this. If the cost-control system developed and adopted becomes too detailed then the cost of controlling costs becomes disproportionate to the benefit derived. Therefore rather than design a system of cost control which considers every item it is better to concentrate on the key elements.

13.3 The basic needs of a cost-control system for the contractor and client

Any cost-control system used by the contractor must ensure payment for work done by having automatic triggers built in. It should be able to monitor costs as the work progresses at least monthly and indicate the amount of losses or gains being made, anticipating any cash-flow problems. There should be limitations in the system that prohibit any changes to the work in excess of a given amount without the authority of senior management. Similarly, certain personnel may have restrictions on the amount they can spend without recourse to a senior person. It should be transparent and understood by all those whom it affects, whether in-house, suppliers or sub-contractors.

The client needs a system that does not permit any alteration to the tender cost of the contract without authorisation by either the client or representatives. The client or representatives should agree any changes made which affect the tender sum above a certain limit before the instruction is given to the contractor. Any likely overspend from the agreed budget should be highlighted as soon as possible to enable the employer to have time to finance the excesses or rein back expenditure to the initial budget. The system should be such that the client can monitor, on a monthly basis, the current financial state of the project. It should be readily understood by the client, representatives and the contractor.

13.4 Cash flow

Cash flow is the transfer of money into and out of a company. As indicated previously, the developer needs to know what financial demands are going to be made during all stages of the development. Equally, contractors need to know about their predicted cash flow to be able to cover any deficit in funding. Remember the contractor aims to cut this to a minimum as the margins on a contract are small and any interest paid on money borrowed will eat into this.

In practice, on a traditional contract using the bills of quantities the timescale is as follows. To save time the contractor's quantity surveyor will produce an interim valuation of the work that has been carried out by the contractor during the month. A meeting then takes place with the client's quantity surveyor (PQS) on site at the start of the following month and this valuation will be agreed. It normally takes one week for the PQS to submit this valuation to the architect. Usually the architect takes two weeks before issuing a certificate, and depending on the clause in the contract the employer is obliged to pay the contractor the agreed amount within 14 days. This can sometimes be 28 days. Generally, from completing the month's work the contractor is paid within four to five weeks.

It used to be that when the contractor received this money, the sub-contractors would then be paid. However, this practice on major works has largely died out, as the sub-contractor is entitled under their contract with the main contractor to receive their payment after a fixed period of time, often 28 days. In practice this means that by and large the main contractor receives payment at the same time as it is necessary to pay the sub-contractor. It is important this latter payment is made, as the sub-contractor has to pay the staff and operatives carrying out the work, many of whom will be paid hourly which means they are paid the week following the one they have worked. Withholding this payment could land the sub-contractor in a financially difficult position.

The tightness of the timescale also reinforces the previous discussion on the importance at the decision on whether or not to tender, by the main contractor, in establishing the financial strength and reliability of the client's payment record.

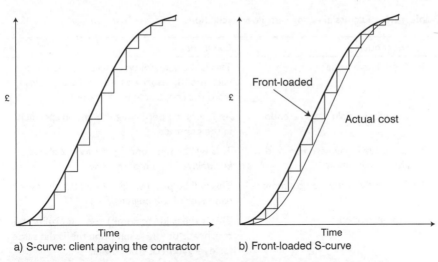

a) S-curve: client paying the contractor   b) Front-loaded S-curve

Figure 13.2 a) S-curve: client paying the contractor; b) front-loaded S-curve

13.5 S-curve

An S-curve is a graph plotting the cumulative income against time. The time scale is normally in monthly increments to coincide with the way contractors are paid by the client.

The S-curve in Figure 13.2a demonstrates what happens when the client pays the contractor. At the end of each month, the contractor submits an interim valuation for the sum of work carried out. However, these monthly valuations will not be paid for another 4 to 5 weeks, shown by the horizontal line on the steps. The vertical line on the steps represents the interim payment from the client. If the payment is made within the month then it can be seen that the main contractor will not have a cash-flow problem. However if the contractor front-loads the tender by a small amount, then as can be seen in the Figure 13.2b, the contractor receives the money from the client well before it is necessary to pay the sub-contractors. The cash-flow deficit probability is then limited to the back end of the contract.

13.6 Further refinements

To carry out a full and detailed calculation, there are many more factors to consider. Adapted from Harris and McCaffer (2001), Table 13.1 shows these factors and indicates how the calculations are derived. These would be carried out at monthly intervals throughout the duration of the contract in a similar way to that used in calculating discount cash flow (section 3.3.3).

Table 13.1   Factors affecting cash-flow calculations

	Description	Comments
1	Cumulative value	This is the value of work to be carried out each month, usually based on the activities from the construction programme
2	Cumulative value less retention	As 1, less the percentage retention specified in the contract
3	Cumulative payment received after certification	This will be the same figures as in 2 above, but delayed by a month plus
4	Cumulative retention payment	This will be paid usually 12 months after the handover of the building
5	Cumulative cost	This is the value (column) less the profits and overheads on the assumption this is the same for every item of work done.
6	Cumulative direct labour costs	These payments are in effect made as and when the work is done. Note comments in 9 below.
7	Cumulative materials costs	Suppliers are usually paid 28 days after delivery, so this cost goes in the following month after receipt of goods.
8	Cumulative plant costs	As 7 above
9	Cumulative sub-contractor costs	Normally 28 days after work is completed paid on a monthly basis, so shown in the following month to when work was valued. Note that in certain cases of labour only sub-contracts this can be as 6 above.
10	Cumulative cash out 6 + 7 + 8 + 9	Direct labour + Materials + Plant + Sub-contractors
11	Cumulative cash flow 3 + 4 – 10	This line becomes the data to demonstrate the cash-flow situation month by month
12	Fixed preliminaries	This is the amount paid up front by the client
13	Term related preliminaries	The monthly outgoings in running the contract
14	Contribution to overheads	This is the difference between 1 and 2 above

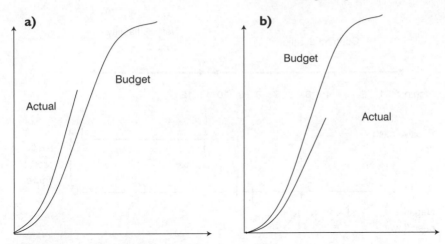

Figure 13.3 Monitoring trends

13.7 Monitoring using the S-curve

Simply by monitoring the actual performance with the predetermined budget, one can see what is happening and the results. Figure 13.3a represents the budget forecast of expenditure and the work actually carried out at a point in time on the project. This graph can be produced for the project costs or sections, such as materials. In this case the trend is positive. Figure 13.3b represents the budget forecast as before, but in this case the actual value of work carried out is less. Any continuing trend in this direction will result in increased financial problems for the company.

13.8 Saw-tooth diagram

Another means of demonstrating the cash flow is by using a saw-tooth diagram. In many ways this is a better way of demonstrating the position so it can be seen what financing is required month by month. The cash flow is shown in linearly, plotting negative and positive cash flows against time.

At the start of the contact as shown in Figure 13.4, the interim payment for the first month's work is received at the end of the second month (approximately) as shown by the vertical line and so on throughout the duration of the project. For simplicity is assumed the expenditure is constant through out each month, although in practice this may not be the case.

In this example, the lowest point of negative cash flow occurs at month six. This is the maximum cash requirement the contractor has to provide to fund the project and it would be important to know this at the tender stage.

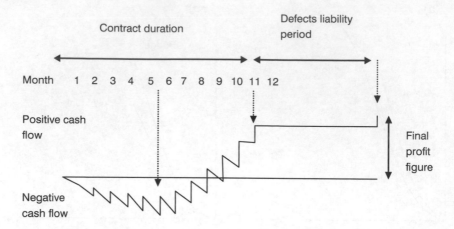

Figure 13.4   Saw-tooth diagram

If during the course of the contract the position worsened because of delays, then this would be continually looked at the see what the implications are.

At month 12, the defects liability period commences, the end of which the retention monies are paid. The final profit released is shown at the end of the defects liability period. It is possible to superimpose the actual position in terms of valuation or cost on this diagram.

13.9 Bad practices

The are a number of questionable practices that have arisen over the last few decades which assist the contractor and sub-contractor to maintain a good cash flow. Since Latham (1994), and subsequently Egan (1998), there has been a trend and pressure to eliminate such practices and be more open with the client and those involved in the design and construction processes. This trend must be continued for the health of the industry especially with the development of supply chain management and partnering initiatives.

Practices to be avoided include:

- front-loading: not changing the total tender price, but increasing the items constructed early and reducing those at the end to balance the overall price;
- back-loading: at times of high inflation, loading he latter items to be constructed at the expense of the earlier ones;

- over measurement: for example agreeing that a floor on a multi-storey reinforced concrete building is 7/8 complete rather than the 3/4 in reality;
- delaying payments to sub-contractors and suppliers until receiving payment for the client; this may incur a loss of discount for not paying within the contracted period for payment, but may still be profitable;
- entering into dispute with the sub-contractor over variations to delay payment;
- disputing the account to delay payment.

The client's representatives can employ similar practices:

- entering into dispute with the contractor over variations for work done
- disputing the accounts;
- delaying passing information to the architect for the issuing of interim certificates;
- delay issuing the interim certificates.

References

Cooke, B. and Jepson, W. (1979) *Cost and Financial Control for Constructions Firms*, Macmillan Press.

Harris, F. and McCaffer, R. (2001) *Modern Construction Management*, 5th edn, Blackwell Science

Heinze, K. (1996) *Cost Management of Capital Projects*, Marcel Dekker Inc.

Potts, K. (1995) *Major Construction Works: Contractual and Financial Management*, Longman

Interim valuations, claims and variations

14.1 Introduction to interim valuations

Interim valuations are for work completed during each month of the contract and are a standard feature of construction forms of contract. What is written into the terms of the contract will determine who is responsible for measuring the work done, but for most building contracts it will be the quantity surveyors (QS) who carry out this role. It is normally the responsibility of the client's QS to measure and value the work, but to expedite the situation the contractor's QS is usually involved. Indeed it would be normal for the contractor's QS to have already measured the work and then just agree it, with some auditing of course, with the client's QS.

Not only do these valuations act as the basis for the contractor and supply chain to receive money for work done, it also acts as a major source of data for any monthly cost-control system, as this is the actual money received and not a theoretical sum.

The valuation for the interim certificate is the gross agreed valuation of work less any amount deducted for retention (section 14.2) and materials on and off the site (sections 14.4 and 14.5).

14.2 Sums subject to retention

The amount of retention will have been agreed between the parties as part of their contract, and is usually 5 to 10 per cent. If the client has a strong belief the contractor will complete the works on time and to the quality specified with minimal remedial work required after the completed building has been handed over then the percentage will be in the lower range. The sums normally included in this calculation are:

- the total value of the work properly executed by the contractor including any agreed variations;
- the total value of the materials and goods delivered to site for incorporation into the building; the contractor cannot obtain all of the materials for the whole contract, have them delivered and expect to be paid for them – there must be a phased delivery unless there are good reasons why this cannot happen (section 14.4);
- the total value of goods off the site, but which are to be included into the building. These are known as 'the listed items' (section 14.5);
- the total amount due to nominated sub-contractors which is also subject to a retention;
- the profit the contractor makes on the work executed by nominated sub-contractors;
- fluctuations if the contract allows for this.

14.3 Sums not subject to retention

While the majority of items have the retention percentage applied to them, there are some payments the contractor has to make that are exempt from this. These are generally costs the contractor has to make that would be unreasonable for them to cover and can include:

- any fees or charges made to local authorities or statutory bodies;
- where the contractor has been instructed by the architect to open up for inspection any covered work, or to test any material or goods;
- if instructed by the architect, any costs incurred in infringing any copyright or patents;
- if instructed by the architect, any added insurance costs;
- any justifiable costs not covered elsewhere in the contract;
- any costs in respect of any restoration, replacement or repair of loss or damage and including the removal and disposal of debris;
- final payment due to the nominated sub-contractor after their agreed defects liability period;
- any increase of costs due to fluctuations on national insurance contributions, levies or taxes;
- monies payable to nominated sub-contractors, which are not subject to retention.

14.4 Materials on site

These are materials delivered and stored on the site the contractor intends, within the foreseeable future, to use in the building. Theoretically when the goods are paid for they become the property of the client. However, there are sometimes conflicts if the supplier's conditions of sale stipulate the goods remain the property of the supplier until the contractor pays for them.

The materials once paid for must not be taken off site without written permission; this also applies to domestic sub-contractors once their materials are paid for. When this payment occurs, a transfer of title to the client from contractor and sub-contractor is made providing this can occur legally depending on ownership at the time.

14.5 Off-site materials

These are usually components, plant and equipment such as air conditioners, boilers, etc., that have been made and are awaiting installation. It could also include other materials such as bricks and blocks, especially if these have been made specific to the contract. The conditions to be met include:

* the item(s) have to be listed by the client and supplied to the contractor in the annex to the contract bills;
* the contractor has to demonstrate that the components, etc., are for the building;
* the listed items are in accordance with the contrac;t
* any material or component held at the suppliers premises, must be set aside or clearly and visibly identified;
* the contractor demonstrates to the client that these goods and materials are adequately insured.

14.6 Introduction to claims

Unfortunately the term 'claims' has become somewhat emotive as a result of contractors 'claiming' for everything possible and exploiting every opportunity and loophole in the contract. It can be argued, with some considerable weight of evidence, that much of the disquiet between developers and contractors has resulted from excessive use of these procedures. This was highlighted in the *Faster Building for Industry* report (1983) and both Latham (1994) and Egan (1998) have identified the same problem. However, within contractual arrangements such as in JCT05 there is provision for justifiable claims.

14.7 Extension of time

An application for an extension of time to the contract does not automatically bring the right to extra payment, as it still may be possible to complete the contract on time by, for example if the delays have occurred on a non-critical item on the programme. The following are typical examples of demonstrable delays to completion time:

- discrepancies in the documents such as the contract drawings and the bills that take time to resolve;
- variations resulting in extra work;
- when the architect issues an instruction to postpone any of the work
- exceptionally adverse weather conditions;
- civil commotion, strikes and lockouts;
- the opening up of work for inspection and then found not to be defective;
- failure by the architect/others to supply information by the agreed dates;
- failure by the client to provide goods for installation/inclusion by the agreed date;
- failure by the client to give ingress or egress by the time agreed in terms of the contract;
- the use or threat of terrorism and/or the activity of the relevant authorities in dealing with the matter.

14.8 Cost centres for claims

When building up a claim there are several cost centres which, when added together, form the basis for the claim. Nedekugri and Rycroft (2000) suggest these are: site establishment costs, head office overheads, visiting head office staff, uneconomical working, uneconomical procurement, loss of profit, acceleration, third party settlements, inflation, financing other cost centres, financing retentions, interest, and cost of producing the claim.

14.8.1 Site establishment costs

These are to do with the time-related costs such as the accommodation, servicing of the same in terms of labour for cleaning and electricity, heating, insurance, etc., and any plant standing idle as a result of a delay. It also includes site supervisory staff and non-productive labour, including attendance to sub-contractors. The claim will normally be based on hire invoices and bills for the accommodation, plant, and for the heating, insurance. etc. The staff and labour costs include the total cost to the contractor of any benefits such as pensions, holidays and health insurance.

14.8.2 Head office overheads

This is a difficult item to value and agree with the client who could well argue that those employed in the head office can in part be employed doing other activities in lieu of supporting the contract.

The simplest way, known as the 'Hudson Formula', is to take the percentage added in the bills, divide by the contract period and multiply by the length of the delay. The weakness in this approach is that relates to the tender, which is based on value, rather than actual costs. Equally, it assumes that the overheads for the contract are constant throughout its running period. There are a variety of other ways of doing the calculation, some more popular than others.

14.8.3 Visiting head office staff

This is the additional time spent by staff visiting the site. It can be argued is taken care of in the general head office overheads and this is valid. However, the contra-argument is that the staff still have to visit and their workload may increase as a result of the claims being made by the sub-contractors as a result of the delay. It can cause conflict between the contractor and the client and is probably best, on balance, not followed through.

14.8.4 Uneconomical working

Any disruption on site causes problems that can be exhibited in a variety of ways. If the disruption is severe, the morale of the workforce takes a dive with resulting loss of productivity, and possible industrial relations problems. However, it is difficult to quantify the cost of the disruption. It is not just the loss of production that costs. If extra work of modest proportions is added to the contract it may mean bringing back a piece of plant, operatives have to go through the learning curve again, and extra costs involved in administering the procurement, etc. Under these circumstances it would be unfair to use the current bills item for this purpose.

14.8.5 Uneconomical procurement

This will not happen very often, but there are circumstances that arise where it might be applicable, especially where large deliveries are involved. The conditions of sale of the supplier may be such that if delays over a certain length of time accrue, storage costs would be charged. Alternatively, it may be necessary for the deliveries to continue resulting in storage problems on site and double handling with possible damage.

14.8.6 Loss of profit

Since profit is usually a percentage based on the cost of the works, it could be argued that the contractor will receive the profit as contracted in any case. The other argument is that delays restrict the ability to earn more profit on potential opportunities. It is difficult to quantify and again, tactically it may be better to ignore it unless the disruption and consequent delay is of significant magnitude to have an impact on the business as a whole.

14.8.7 Acceleration

In spite of the delay, the client may still require the project to complete by the original date. A good example of this is in retail. If the project was due to be handed over in time for the retailer to capitalise on the Christmas period, the costs of the delay may well outstrip paying the contractor extra to accelerate production to maintain this date. The client offering financial inducements normally achieves this. Alternatively, the contractor is paid for the extra costs incurred in meeting the client's revised requirements, although there are some legal risks in this. Additionally, the client can persuade the architect to deny extensions or the contractor will take it upon itself to meet the original deadline, without passing on extra costs, as a marketing tool for future work and to enhance its reputation.

14.8.8 Third-party settlements

In today's marketplace, when a project is delayed, because so much of the work is sub-contracted, there is an immediate impact on them. This means that the financial liabilities traditionally borne by the contractor are passed to the sub-contractors. The problem arises in that the negotiation between the contractor and the client over extra costs can take a considerable time. In the meantime the sub-contractors want their money and may well be financially distressed if settlement is not made in a reasonable time. In these cases the contractor may decide to settle with the sub-contractor before agreeing terms with the client.

14.8.9 Inflation

If the contract allows for fluctuations then the comments in this section do not apply, as inflation will automatically be taken into account in the sums. However, if the contract is a non-fluctuating cost project and the client causes delays, in principle, the contractor is entitled to be reimbursed for increased costs due to inflation. The calculation would be based on the differences in start times of activities as compared with the original programme. Again, whether or not the contractor wished to invoke this depends on the extent of the delay.

14.8.10 Financing other cost centres

When borrowing rates from the banks and financial institutions are high, then the costs of financing delays reflect this. If the delays occur at a time when the contractor is running at a planned negative cash flow (section 13.4) then the impact is even greater. However, even when at a positive cash flow it still costs the contractor money, as any shortfall of profit cannot be re-invested at this now higher interest rate.

14.8.11 Financing retentions

When tendering for a contract, the contractor produces a cost-flow analysis indicating the short falls in cash flow. A cash flow for the whole company's operation will also be produced which might, over the period of the financial year, eliminate the need to finance any negative cash flow as the profits from other contracts cover this deficit. The contractor will have made arrangements accordingly. If a contract is delayed there is a similar delay in the payment of the retention monies, which may either cause negative flow, and hence the need for financing, or it lowers the positive cash flow, reducing the amount of potential interest earned from investment. In both cases this would be a legitimate claim.

14.8.12 Interest for non-payment

When payments from the client are paid late the contractor has a right to simple interest.

14.8.13 Cost of producing a claim

Putting a claim together costs money in terms of the man-hours needed to collect the information and make the case. In practice, claims of this nature are made for breach of contract against the client. However, as before it is a matter for management to decide what the long-term implications are to the relationships with the client and the architect.

14.9 Introduction to variations

Variations to the works tendered for are almost certain to occur for some reason or other. Ideally, a building would be designed so this would not happen. This would remove the flexibility, useful to the client, especially since there may be a long gap from the time of the feasibility study to the day the contractor arrives on site. In the meantime the client requirements can change.

Typical reasons for variations are:

- unexpected ground conditions resulting in extra excavation or foundation design changes;
- a change of requirement by the client;
- a design error; when considering the complexity of a modern building it is almost inevitable that errors will occur.

The calculation for variations is beset with potential points of disagreement between the client, the contractor and sub-contractors. It can involve considerable sums of money. The more thorough the company's cost-control documentation, the easier is it to provide evidence to demonstrate the validity of the claim.

The JCT05 contracts clearly lay down a series of conditions that contribute to a variation and lay down the rules for claiming variations. The reader should refer to these documents for further details. If the claim can be measured using a bill item, and the variation agreed there will be no dispute. However, if the claim is of a higher magnitude, how the calculation is considered is an issue. It should be transparent and honest.

14.10 Example calculations

An item in the bills is to excavate a 3 metre deep basement over an area of 25m × 40m. The calculation to establish to bills rate could be as follows:

Stage 1: calculate the cost of excavation

Table 14.1   Excavation costs

Bills quantity 3000m³				
Labour	Rate	Plant	Rate	Total
		excavator	£30/hr (inc. driver)	£30
Ganger	£7/hr			£ 7
2 no.	£6.5/hr			£13
			Total cost	£50
Output rate from excavator is 20m³/hr				
Cost of excavation per m³ is £50/20				£2.50

Stage 2: calculate the cost of removing the excavated soils and cart away to landfill site

The 3000m³ has to be carted away using 8m³ capacity lorries. It takes 24 minutes to load the lorries and the round trip journey to the landfill site is 30 minutes including tipping time. This will require three lorries for the following reasons. Ideally the excavator needs to work continuously to keep the production on the site moving forward. Table 14.2 demonstrates the cycles of excavator and the wagons required to make this happen. It means that each wagon will have to wait for 12 minutes on return from the landfill site. This is not a major problem as it would be necessary in practice to build in some slack to account for variable traffic, delays at the tip and natural breaks for the drivers. Increasing or decreasing the capacity of the excavator can also adjust the balance between the excavator and the wagons, but it must be remembered that the size of the excavator will have been chosen to suit the programme requirements. On the other hand when value engineering the project, an increase in the excavation duration may be one of the ways of achieving a cost reduction, in which case the balance between excavator and lorries can be reconsidered.

The cost of the transportation per metre cubed is therefore based on the decision to use three wagons: 3 lorries cost £60 in total remove 20m³ in one hour, therefore the cost per m³ is £60/20 = £3.00

Table 14.2   Ratio of lorries to excavators

		24 mins	24 mins	24 mins	24 mins	24 mins	24 mins
Excavator	20m³/hr						
Lorry 1	8m³ capacity	Load		Transport		Load	Transport
Lorry 2	8m³ capacity		Load	Transport		Load	Transport
Lorry 3	8m³ capacity			Load	Transport		Load

Stage 3: calculate the bills item

This is shown in Table 14.3 giving a cost/m³ of £6.49 to be used in the bills.

Table 14.3   Bills item calculation

Description	Quantity	Cost/unit measure	Cost
Excavation	3000m³	£2.50	£7,500
Transport to tip	3000m³	£3.00	£9,000
		Total cost	£16,500
Site overheads	10%		£1,650
Profit and overheads	8%		£1,320
		Gross cost	£19,470
		Cost/m³ (divide by 3000)	£6.49

Now assume that the excavation depth has to be increased by 2m as a result of discovering a layer of peat over a major proportion of the site area, but unfortunately was missed by the site investigation. The calculation may now change because the excavator previously selected cannot excavate to this depth. In this case the calculation will be as follows:

Stage 1: calculate the cost of excavation

Table 14.4   Excavation costs

Revised quantity 5000m³				
Labour	Rate	Plant	Rate	Total
		excavator	£40/hr (inc. driver)	£40
Ganger	£7/hr			£ 7
2 no.	£6.5/hr			£13
			Total cost	£60

Output rate from excavator is 32m³/hr

Cost of excavation per m³ is £60/32	£1.875

Note that the cost of excavation in this case shown in Table 14.4 is cheaper than the previous scenario by £0.75/m³.

Stage 2: calculate the cost of removing the excavated soils and cart away to landfill site

The same size lorries are employed taking the same time to transport and tip the soil, but the volume has increased to 5000m³.

Table 14.5 Ratio of lorries to excavator

		15 mins	15 mins	15 mins	15 mins	15 mins	15 mins
Excavator	32m³/hr						
Lorry 1	8m³ capacity	Load	Transport		Load	Transport	
Lorry 2	8m³ capacity		Load	Transport		Load	Transport
Lorry 3	8m³ capacity			Load	Transport		Load

In this example the transportation works out that still only three wagons are required, but the waiting time has been eliminated on return. If this number of lorries is adopted, there could be hold ups in the excavation from time to time due to the possible delays to the lorries as indicated before, so it would become a management decision to decide whether to have an extra lorry or not. The cost of the transportation per metre cubed using the three lorries is £60 in total to remove 32m³ in one hour is £60/32 equating to £1.875.

Stage 3: is to calculate the revised rate

Table 14.6 Revised bills rate calculation

Description	Quantity	Cost/unit measure	Cost
Excavation	5000m³	£1.875	£9,375
Transport to tip	5000m³	£1.875	£9,375
		Total cost	£18,750
Site overheads	10%		£1,875
Profit and overheads	8%		£1,500
		Gross cost	£22,125
		Cost/m³ (divide by 5000)	£4.42

Note that the rate in this case shown in Table 14.6 is less than that in the bills that means that pro rata the client is paying less per m³ although in real terms is paying the difference between that in the bills of £19,470 and the revised price of £22,125.

Had the travel time been, say, 48 minutes, then the situation would have been different again as in the first instance the lorries usage would have been 100 per cent and in the latter, less as shown in Tables 14.7 and 14.8.

Table 14.7   Ratio of lorries to 20m³ excavator

		24 mins	24 mins	24 mins	24 mins	24 mins	24 mins
Excavator	20m³/hr						
Lorry 1	8m³ capacity	Load	Transport		Load	Transport	
Lorry 2	8m³ capacity		Load	Transport		Load	Transport
Lorry 3	8m³ capacity			Load	Transport		Load

Table 14.8   Ratio of lorries to 32m³ excavator

		15 mins	15 mins	15 mins	15 mins	15 mins	15 mins
Excavator	32m³/hr						
Lorry 1	8m³ capacity	Load	Transport				Load
Lorry 2	8m³ capacity		Load	Transport			
Lorry 3	8m³ capacity			Load	Transport		
Lorry 4	8m³ capacity				Load	Transport	
Lorry 5	8m³ capacity					Load	Transport

For the 3000m³ excavation the comparative costs would be as before, £6.46. In the case of the 5000m³ excavation, the cost of the excavation is the same at £1,875/m³, but the cost of the transportation has increased to 5 lorries costing £100 in total to remove 32m³ in one hour making a cost per m³ of £100/32 equating to £3.13.

Table 14.9 demonstrates the revised calculation and it can be seen that whilst more expensive than shown in Table 14.6 it is still cheaper per m³ than the original bills item. Another solution would be to reduce the number of wagons on the 5000m³ excavation and let the excavator wait the three minutes to allow the wagons to return.

Mention was made earlier in this section to using this information for value engineering, and altering the size of the excavator can reduce the cost of the operation. From the examples it can be seen that in the case of the

Table 14.9   Revised cost for 32m³ excavator

Description	Quantity	Cost/unit measure	Cost
Excavation	5000m³	£1.875	£9,375
Transport to tip	5000m³	£3.13	£15,550
		Total cost	£24,925
Site overheads	10%		£2,492
Profit and overheads	8%		£1,994
		Gross cost	£29,411
		Cost/m³ (divide by 5000)	£5.88

shorter haulage distance and the longer one, the use of the larger excavator reduces the unit cost of excavation and carting to tip rate. If the excavation remained at the initial volume (3000m³) it would be cheaper to use the larger excavator. The effect also in this case would be to reduce the time for the excavation from 3000/20 = 150 hours to 3000/32 = 94 hours, i.e. from four weeks to just over 2.5 weeks.

Regardless, what is demonstrated in these examples is as the parameters vary, there will be an effect on the cost of the variation.

References

JCT05 Design and Build Contract (2005) RIBA Publications.
JCT05 Standard Form of Building Contract with Quantities (2005) RIBA Publications
Ndekugri, I. and Rycroft, M. (2000) *The JCT98 Building Contract: Law and Administration*, Arnold.

15

Cost systems

15.1 Introduction

The purpose of a cost-control system is to ensure costs are controlled so that the contract is completed within budget and lessons are learnt through appropriate feedback. It is important to note this means investigating not just the causes of loss-making activities, but also those which produce a high rate of return. The latter because in today's market environment the price of similar work in the future might be reduced making the tender more competitive. This chapter looks at contractor cost-control systems rather than those of the client.

Traditionally, the industry used to record costs and compare these with the budget (section 15.2). The problem with this approach is it provides historical information when it may be too late to do anything about it on the current contract, although it does provide feedback after analysis for the future. However, where there is repetitive work the information can be analysed and used for the rest of the contract. For example, it was established that the foundation slab of the first of four 16-storey blocks of flats had cost a considerable amount of money compared to the budget. On analysis it was concluded the main cause was the method adopted. By this time, the second block's foundations were too far advanced to change, but a revised method was adopted for blocks three and four.

Ideally what is needed a cost-control system that is cost forecasting rather than cost recording.

15.2 Budget cost

The budget cost can be derived from a variety of sources depending on the form of contract and accompanying documentation. For example, with a traditional bills of quantities, the cost of each item is derived from the build up of cost elements of plant, materials and labour, so budgets can easily be produced for either a unit measure, for example, a metre run of skirting, or the cost of a larger element such as a complete storey of reinforced concrete including walls, columns, beams and floors.

Another way of establishing budgets is to cost activities on the programme then assess – at the time of the cost analysis – what percentage of the activity has been completed. The advantage of this approach is that there are programme management packages, which will produce costs based upon the programme constructed.

With the industry using a high percentage of sub-contractors, the budgets are based on the sub-contractor's price quoted for the contract. The work done by them can be calculated as a percentage of their work completed and the materials consumed in the process.

15.3 Actual cost

For directly employed labour this can be quite a complicated process depending on the system in use. The following considerations need to be taken into account:

- each week the employees are paid for the hours they have worked and a bonus based on their productivity;
- the bonus is usually paid one week later than their wage as it takes a week to calculate;
- the wage might also include guaranteed hours and bonus due to, say, inclement weather;
- employer's contribution to national insurance;
- employer's contribution to pension, in some cases;
- trades employees are paid a different rate to general operatives;
- wages may include travelling costs;
- plus rates in lieu of bonus payments;
- plus rates for tool money;
- holiday pay.

The budget labour figure from the bills of quantities includes for other items (itemised in section 10.8) used to build up the tender rate and also accounts for the summer and winter working hours to provide a wage rate to be used in the estimate.

Reconciliation between these two sets of information and figures is not as easy as it seems. This means either the actual cost has to be adjusted starting with the wage sheet information, deducting the bonus paid and adding the previous wage sheets bonus, and then adding a figure which accurately reflects the other cost items not shown on the wage sheet or, the budget figure is amended by deducting all the cost items that do not appear on the wage sheet. In the latter case, these costs have to be reflected somewhere else in the overall balance of costs.

It is probably simpler to adopt the former approach. It is necessary to calculate the total number of hours for which the general operatives and the tradesmen are paid, rather than the number of hours they have worked (shown on the left side of Table 15.1) and calculate the total wage each type of labour has received in total (as shown on the right side).

By dividing the total of the wages columns by the total of the hours column, an hourly rate can be calculated for general operatives and tradesmen.

When dealing with sub-contractors' actual costs, the matter is much simpler as one is paying for the work that has been completed and this is as per their quotation to the contractor. The monies that contribute to the contractor's account would include any attendance allowance or profit and overhead associated with their work.

Table 15.1   Weekly labour costs

General operatives (GO)			
Breakdown of hours		Breakdown of wages	
Normal working hours		Flat rate wages	
Overtime hours worked		Plus rates	
Non-productive overtime		Bonus (previous week)	
Travelling time		Tool money	
Maintenance time		Expenses	
Inclement weather time		Lodging allowance	
Guaranteed make-up time		National Insurance contributions	
		Holiday pay	
		Pension contributions	
Total	GO hours		GO wages
Tradesmen (TM)			
Hours as above	TM hours	Wages as above	TM wages

Note:
GO + TM hours should equate to the total hours on wage sheet.
GO + TM wages should equate to the total monies paid out taking account of the bonus adjustment and any costs not actually paid directly to the operatives.

15.4 Direct labour costing systems

15.4.1 Historical

These systems are based on the accountancy principle of recording every thing in detail and producing accounts later that demonstrate what has happened. It was a very slow and laborious process. It used to be assumed that one cost surveyor was needed to calculate the bonus and costs for 100 operatives and the operation would take a full working week. The reason for discussing it here is that there is a belief that with the advances made in information technology, possible developments in software and the use of pointers able to measure dimensions from computer-aided drawings, there is no reason why this approach should not be resurrected, providing the opportunity for some manual intervention is permitted. If this was developed then the operation could be completed in one or two days and management would have much more up-to-date information with which to make decisions especially for large sub-contractors who employ labout directly.

The first stage is to collect information about the work that has been carried out by each person or gang of operatives and how long they have taken to carry out the work, including any non-productive time. This is provided by the supervisor daily or weekly on a time sheet. This information is then abstracted and entered on an allocation sheet (Tables 15.2 and 15.4).

On the left side of the page as shown in Table 15.2 the details of the operatives and how much time they have spent working on the contract during the week is entered. It can also be used for the bonus calculation.

Table 15.2 Actual hours worked

Allocation sheet

Week commencing:

Operative	Works no.	M	T	W	Th	F	Total	Bonus
Jones	1345	9	9	8.75	9	8	43.75	£75.00
Smith	1243	8.5	9	8.75	8.75	8	43.00	£74.00
							86.75	£149.00

Bonus calculation
Hours earned = 124
Actual hours = 86.75
Hours gained 37.25 @ £4/hr = £149
£149/86.75 = £1.72 per hour

The times allocated against the gang are those the operatives have actually worked, based on the contractor's system of recording entry on and off the site, such as clock cards. In practice these may be different from the time sheets provided by, or on behalf of, the gang (Table 15.3), where it is not unusual for the hours to be rounded up to the nearest half and ignoring the fact that the operatives may have been quartered. Quartering is when the operative has clocked in excess of three minutes past each quarter. The operative will be paid as if having commenced at the start of the next quarter period.

A typical daily sheet for Monday is shown in Table 15.3. In this example the two second-fix joiners have allocated a total of 18 hours, yet according to the times booked into and out of the site as shown in Table 15.2 they have only been recorded for a total of 17.5 hours.

The right side of the allocation sheet, shown in Table 15.4, is used to allocate the work and is taken from the time sheets. This has to account for the actual hours worked rather than that submitted on the time sheets. It is usual practice to make these minor adjustments in either the main non-productive items, or if none are identified, from those productive items with the largest amounts of time allocated. This adjustment is made on 'await instructions' shown in italics in the Monday column in Table 15.4.

The total hours for the gang when added up should be reconciled with the total hours worked by the gang as shown on Table 15.2, i.e. 86.75 hours.

The quantity and descriptions of the work done, usually separated into productive and non-productive, work is then entered. The column 'Rate' is

Table 15.3 Daily time sheet

Daily worksheet	Names	Works no.	Trade
Date:	Jones	1345	Joiner
	Smith	1243	Joiner
Operation	Quantity	Hours worked	
Productive			
Fix doors	10 No.	7 hours	
Fix skirting	22m run	4 hours	
Fix architrave	16m run	3 hours	
Non-productive			
Await instructions		2 hours	
Replace broken door		2 hours	
	Total	18 hours	

Table 15.4 Work allocation

Description	Quantity	Rate	Hours earned	M	T	W	Th	F	Total
Productive work									
Fix doors	26	1.5hr/no.	39	7.0	5.0	3.0	5.5	3.0	23.5
Fix skirtings	93	3m/hr	31	4.0	2.0	5.0	4.0	4.5	19.5
Fix architraves	120	3m/hr	40	3.0	7.0	5.0	5.0	6.0	26.0
Fix window cills	6	2hr/no.	12		2.5	3.0	3.25		8.75
Fix notice board	1	2hr/no.	2					2.0	2.0
Non- productive									
Await instructions			0	*1.5*				0.5	2.0
Replace broken door			0	2.0					2.0
Clean up			0		1.5				1.5
Re-fix skirting			0			1.5			1.5
Total			124	17.5	18.0	17.5	17.75	16.0	86.75

the bonus rate agreed between the company and the operatives and from this the bonus hours earned for each of the described activities can be calculated, giving a total, in this example, of 124 hours. The bonus calculation is relatively simple as is shown on Table 15.2 where the hours worked are deducted from the hours earned and this is then multiplied by whatever rate has been agreed between the parties. In this case £4 using a 50 per cent scheme (*Business Organisation for Construction*, Chapter 8) for a total bonus for the gang of £149 that is pro-rated for each member of the gang depending on the hours worked. In this calculation, the non-productive items are given no earnings at all. This can be a cause of dispute, since it can be logically argued that if, for example in the case of the 1.5 hours 'awaiting instructions', it is management that is preventing the operatives from earning a bonus they should be given the time as a contribution to the bonus or be paid pro-rata to the bonus paid.

The quantity and hours spent fixing, in this case skirting, are then transferred from all the appropriate allocation sheets to a cost ledger sheet, as shown in Table 15.5, so a weekly total of this item can be ascertained, as can a running total for the contract. The weekly rate is calculated from Table

Table 15.5   Cost ledger: fix skirting

Cost ledger – second-fix joinery

Sub set

Fix skirting			Labour budget rate £X per metre (derived from the bill)					
Date	Quantity	Hours	Weekly rate	Actual cost	Budget cost	Loss	Gain	Cumulative
Wk Com.	metre		£	£	£	£	£	£

Table 15.6   Weekly summary for second-fix joinery

Cost ledger – weekly account

Second-fix joinery

Description	Actual cost (£)	Budget cost (£)	Gain (£)	Loss (£)
Doors				
Skirtings				
Architraves				
Notice boards				
Cills				
Window ironmongery				
Door ironmongery				
Etc.				
	Total			
	Profit/Loss			

15.1, and thus the actual cost (weekly rate × hours) can be deduced. The budget cost comes from the bills and with this the loss or gain for the week along with a cumulative position can be calculated for the contract for this particular task, that of fixing skirting.

This information can then be transferred to a weekly account for the second-fix joinery, as shown in Table 15.6, and to a weekly summary for all the work carried out on the contract that week (Table 15.7) and finally to a monthly statement (Table 15.8).

The weekly summary is used by site management as a control tool, with the monthly being sent to the regional or head office to advise on how the site is progressing and to help with the overall financial position of work in progress.

Table 15.7   Weekly summary for the contract

Cost ledger – weekly summary				
Description	Actual cost (£)	Budget cost (£)	Gain (£)	Loss (£)
Superstructure				
Cladding				
First-fix joinery				
Second-fix joinery				
Partitions				
Etc.				
	Total			
	Profit/Loss			

Table 15.8   Monthly summary for contract

Cost ledger – monthly summary				
Description	Actual cost (£)	Budget cost (£)	Gain (£)	Loss (£)
Superstructure				
Cladding				
First-fix joinery				
Second-fix joinery				
Partitions				
Etc.				
	Total			
	Profit/Loss			

15.4.2 Forecasting/predicting

These are systems that rapidly anticipate future problems, especially on matters concerning cash flow. Figure 15.1 shows a network diagram of a small project which converts into a bar line as shown in Figure 15.2. How to produce a programme can be seen in *Operations Management for Construction*, Chapter 2.

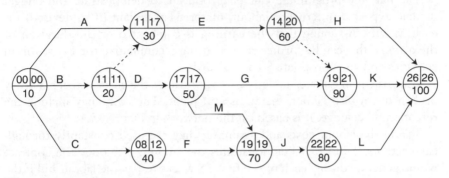

Figure 15.1   Network diagram

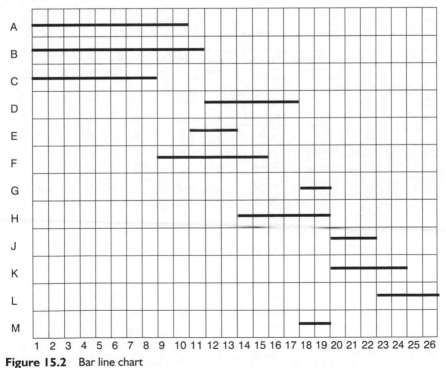

Figure 15.2   Bar line chart

The number in the top left of each circle is the earliest time an activity can finish, in the right the latest it can start and still complete the programme on time and the number in the bottom is used to identify an activity. The programme management package used to produce these programmes needs this, so for example activity A is defined as 10-30. A (10), B (11), C (8), D (6), E (3), F (7), G (2), H (6), J (3), K (5), L (4), M (2).

The bar line programme can be produced to demonstrate the budget income expected in terms of labour, materials and plant (if so devised) for each activity. Assuming that the earning is constant over the duration of the activity, this can be further refined using a cumulative total cost for all activities and an appropriate S-curve drawn.

Using the programme, Table 15.9 demonstrates the costs for each activity broken down into labour, materials and plant. The costs may include the retention percentage. It is based on the network in Figure 15.1.

Taking the labour costs and assuming they are used constantly through each activity, Table 15.10 demonstrates the amount of money the contract would generate during each of the first 15 weeks to pay the labour bill if the programme is adhered to. The sums are rounded to the first decimal place.

Table 15.9 Breakdown of costs per activity

Activity	Activity	Duration	Predecessor	Gang size	Labour cost (£K)	Material cost (£K)	Plant cost (£K)	Total cost (£K)
Description	Comp No.							
A	10–30	10	Start	4	40	37	8	85
B	10–20	11	Start	3	31	30	2	63
C	10–40	8	Start	5	37	30	0	67
D	20–50	6	10–20	6	36	25	0	61
	20–30	dummy	10–20	0	0	0	0	0
E	30–60	3	10–30, 20–30	3	8	9	1	18
F	40–70	7	10–40	5	33	38	7	78
G	50–90	2	20–50	6	11	10	0	21
	60–90	dummy	30–60	0	0	0	0	0
H	60–100	6	30–60	3	15	15	0	30
J	70–80	3	50–70, 40–70	5	15	18	4	37
K	90–100	5	60–90, 50–90	6	28	25	6	59
L	80–100	4	70–80	4	16	17	0	33
M	50–70	2	20–50	3	5	4	0	9

Table 15.10 Breakdown on weekly labour budget costs

Activity	£K	£K	£K	£K	£K	£K	£K	£K	£K	£K	£K	£K	£K	£K	£K
A	4.0	4.0	4.0	4.0	4.0	4.0	4.0	4.0	4.0	4.0					
B	2.8	2.8	2.8	2.8	2.8	2.8	2.8	2.8	2.8	2.8	2.8				
C	4.6	4.6	4.6	4.6	4.6	4.6	4.6	4.6							
D												6.0	6.0	6.0	6.0
E											2.8	2.7	2.7		
F									4.7	4.7	4.7	4.7	4.7	4.7	4.7
G															
H														2.5	2.5
J															
K															
L															
M															
Total	11.4	11.4	11.4	11.4	11.4	11.4	11.4	11.4	11.5	11.5	10.3	13.4	13.4	13.2	13.2
Weeks	1	2	3	4	5	6	7	8	9	10	11	12	13	14	15

15.4.3 Package costing

Now that so much work is let out in packages a further method of controlling costs has evolved which is used particularly in management contacting and construction management contracts where the management team has a clear responsibility to ensure the building is built to budget by controlling design cost implications and construction costs.

The client's quantity surveyor produces a cost plan of the design that is modified and agreed by all parties. This then acts as the cost-control document. The management contractor then allocates these costs to the various activities to be provided by suppliers and sub-contractors. These are referred to as packages. When the managing contractor obtains quotations, hopefully the tender prices equate to those in the cost plan. If not, then negotiation will take place to try narrow or close the gap. If the price in the plan cannot be met, then there are various options:

- the deficit can be balanced against other packages that may come in under budget;
- there can be a modification to the specification to bring it in line;
- a new method can be value engineered.

As the project progresses, any increases in the cost of a package caused by variations or claims are immediately noted and the effect analysed so that the three options above can be considered. As all this information is put on a spreadsheet, the cost impact can be seen immediately so that action can be taken.

References

Griffith, A. and Watson, P. (2004) *Construction Management*, Palgrave Macmillan.
Harris, F. and McCaffer, R. (2006) *Modern Construction Management*, 6th edn, Blackwell.
Pilcher, R. (1992) *Principles of Construction Management*, 3rd edn, McGraw-Hill.

Index